Securing the Current: How Systems, Societies and Institutions Shape the Reliability of the Electric Grid

Exploring the External Forces Surrounding the Grid, From Governance and Affordability to Social Cohesion and Global Stability

Published 2026 by River Publishers
River Publishers
Broagervej 10, 9260 Gistrup, Denmark
www.riverpublishers.com

Distributed exclusively by Routledge
605 Third Avenue, New York, NY 10017, USA
4 Park Square, Milton Park, Abingdon, Oxon OX14 4RN

Securing the Current: How Systems, Societies and Institutions Shape the Reliability of the Electric Grid: Exploring the External Forces Surrounding the Grid, From Governance and Affordability to Social Cohesion and Global Stability / Sofia Weir.

Routledge is an imprint of the Taylor & Francis Group, an informa business

ISBN 978-87-438-1347-7 (paperback)
ISBN 978-87-438-1350-7 (online)
ISBN 978-8-743-81349-1 (ebook master)
ISBN 978-87-438-1352-1 (ePub)

DOI: 10.1201/9788743813491

Securing the Current: How Systems, Societies and Institutions Shape the Reliability of the Electric Grid

Exploring the External Forces Surrounding the Grid, From Governance and Affordability to Social Cohesion and Global Stability

Sofia Weir
EIT, USA

River Publishers

Routledge
Taylor & Francis Group
NEW YORK AND LONDON

To my husband—I have never known a person to be so self-sacrificing for someone else's dream; you alone enabled me to write this book by providing me with the space, compassion, and support I needed to do so in the face of a mountain of familial responsibilities.

Contents

Preface

Reliability has traditionally been treated as an engineering property, defined through adequacy, stability, and compliance. Yet beneath these measurable indicators lies something more fundamental: the relationship between a society and the infrastructures that sustain it. The electric grid, more than any other system, embodies that relationship; it is a technical organism animated by social order, economic stability, and institutional trust.

This book was born from the realization that the reliability problem is no longer purely technical. Despite decades of standards and operational discipline, major disruptions continue to occur not because we misunderstand the grid, but because we underestimate the influence of the world around it. Economic volatility, governance quality, affordability, social cohesion, and geopolitical risk now shape reliability as decisively as physical condition or operator performance. The central argument is simple: reliability is contextual. A grid performs only as well as the environment that enables it. Understanding reliability therefore requires understanding the systems—political, institutional, and human—that surround it.

Each chapter provides an increasingly specific view of reliability, beginning at the global understanding of reliability, followed by a chapter on disaster exposure, affordability, and institutional design, then becoming even more specific eventually closing in on the influential role of geopolitical risks and security on grid reliability before synthesizing the text in the closing chapter. Throughout, you'll find several "undercurrent sections", some briefer than others. These passages serve as reflections on contextual forces that run beneath the main analytical narrative, much like the unseen undercurrents that shape the flow of a river. Each undercurrent captures a foundational idea, historical echo, or societal tension that influences reliability from below the surface, ensuring that the reader remains connected to the deeper themes guiding the reliability discussion. You'll also find textbook-style exercises at the close of each chapter, which will enhance learning for the casual reader and student alike. For additional materials, such as teaching aids, slide decks/presentations, or other information for use in a classroom-setting, reach out to the author.

This is not a prescriptive text. It offers no new standard and critiques no existing institution. Its purpose is to expand the intellectual territory of reliability by integrating perspectives from engineering, systems theory, and the social sciences. It invites collaboration across domains that have long addressed the same challenges from different directions.

What follows is an argument for integration. Reliability should not be seen as an engineering silo but as a shared condition of modern civilization. To study it is to study the continuity of order itself: the capacity of complex societies to sustain the ordinary amid extraordinary change.

Acknowledgements

This work would not have been possible without the many individuals who have crossed my professional and personal path who challenged my thinking, growing my capabilities and sharpening my analysis, including teachers, advisors, friends, and colleagues across my multiple lines of work to date.

On a personal note, I am deeply grateful to my friends and family who tolerated and continue to tolerate my obsessions, and Ross Johnson who not only contributed a chapter foreword, but provided much needed guidance and encouragement early on in both personal and professional capacities. A special mention goes to my colleague Sue Clifton whom I met at just at right time, who refreshed my drive and gave me just the push I needed to get to the finish line.

I'm thankful to the publishers for their work on this manuscript, and for extending this wonderful opportunity to me. Any errors or omissions here are of course solely my own.

Finally, I am thankful to the many operators, engineers, analysts, planners, coordinators, policymakers, field personnel and more, whose daily work keeps critical infrastructure marching on. This book is written in recognition of their efforts and in the hopeful service of more reliable grids across the globe.

Disclaimer: This publication is an independent work that reflects the research, analysis, and perspectives of the author in a personal capacity. It does not represent the views, positions, policies, or interpretations of the North American Electric Reliability Corporation (NERC) or the Electricity Information Sharing and Analysis Center (E-ISAC). Nothing in this work should be construed as commenting on, interpreting, modifying, or evaluating NERC

Reliability Standards, ERO Enterprise activities, or any regulatory authority or jurisdiction. Any references to bulk power system reliability, institutional structures, or comparative international contexts are based on publicly available information and general sector analysis. All conclusions, interpretations, and any errors are solely those of the author.

About the Author

Sofia Weir is an energy-sector analyst and reliability professional whose work explores the intersection of infrastructure resilience, policy design, and societal stability. She serves with the Electricity Information Sharing and Analysis Center (E-ISAC), part of the North American Electric Reliability Corporation (NERC), where her focus includes physical threat intelligence, grid reliability, and the external factors that shape security and system performance.

Before joining the power sector, Weir founded and operated a commercial construction company in Dallas and later served as a facilities engineer at the U.S. Supreme Court. These experiences gave her a grounded understanding of the built environment and the human systems that sustain it.

She holds a B.Sc. in Mechanical and Energy Engineering and an M.Sc. in Energy Policy and Climate from Johns Hopkins University, and is certified as an Engineer-in-Training (E.I.T.). Weir is an active member of the Institute of Electrical and Electronics Engineers (IEEE), participating in the Power & Energy Society, Reliability Society, and Systems Council. Her professional and research interests center on how economic, social, and environmental dynamics influence the reliability of electric systems—a theme first explored in her graduate research on Venezuela's grid collapse.

Trilingual in English, Portuguese, and Spanish, she often integrates cross-cultural perspectives into her work on resilience and energy access. She is also the founder of the Big Horn Energy Equity Group, a nonprofit dedicated to addressing regional energy inequity and access challenges across the Mountain West.

When not writing or working in reliability analysis, Weir enjoys time with her family in Wyoming, making music, and pursuing creative projects that combine art, science, and public service. Her career reflects an enduring belief that reliable electricity underpins not just modern economies, but human dignity itself.

To reach Sofia, find her on LinkedIn or shoot her an email: sofiaweir@gmail.com.

List of Abbreviations

BES – Bulk electric system

BPS – Bulk power system

CAIDI – Customer average interruption duration index

CAIFI – Customer average interruption frequency index

CSE – Continental synchronous area

E-ISAC – Electricity Information Sharing and Analysis Center

E.I.T. – Engineer-in-Training

ENTSO-E – European Network of Transmission System Operators for Electricity

ERO – Electric Reliability Organization

FEMA – Federal Emergency Management Agency

FERC – Federal Energy Regulatory Commission

FFP – Fund for Peace

FSI – Fragile states index

GAO – Government Accountability Office

GDP – Gross domestic product

IEEE – Institute of Electrical and Electronics Engineers

ISO – Independent system operator

kWh – Kilowatt-hour

MED – Major event day

NERC – North American Electric Reliability Corporation

NGO – Non-governmental organization

NOAA – National Oceanic and Atmospheric Administration

PJM – PJM Interconnection, LLC (regional transmission organization)

SAIDI – System average interruption duration index

SAIFI – System average interruption frequency index

SCADA – Supervisory control and data acquisition

SDG – Sustainable development goal

SoS – System-of-systems

SOR – State of reliability (report)

UN – United Nations

UNDP – United Nations Development Programme

UNESCO – United Nations Educational, Scientific and Cultural Organization

WHO – World Health Organization

CHAPTER

1

Beyond Access: Reliability in a Global Context

Abstract

Electricity is foundational to modern life, enabling health, education, economic development, communication, and social stability. Yet nearly 800 million people still live without access to electricity, and millions more experience unreliable or unsafe supply. While global electrification initiatives have successfully expanded access, they have often done so without equal attention to the conditions required for reliable electricity delivery. Reliability is not simply a technical characteristic of grid operation; it is shaped by the social, economic, institutional, and environmental contexts in which electric systems exist. In highly developed regions, reliability planning focuses on operational stability and bulk power system adequacy. In underdeveloped or unstable environments, however, reliability challenges are intertwined with governance quality, resource constraints, security risks, and infrastructure interdependencies. This chapter expands the understanding of electric grid reliability beyond internal system dynamics to include the external pressures that influence whether electricity can be delivered consistently, equitably, and safely, even in the most challenging environments.

DOI: 10.1201/9788743813491-1

1.1 Introduction

Electricity is more than a utility; it is a precondition for human flourishing. In every society where electricity becomes dependable, life expectancy rises, educational attainment expands, food security improves, and economic productivity accelerates. But access to electricity alone is not enough. Electricity must be reliable: available when needed, of sufficient quality to power daily life, and resilient to the disruptions that otherwise plunge communities back into darkness. When electricity flickers, so do the institutions and livelihoods that depend on it. The distinction between electrification and reliable electrification is therefore not a technical nuance, it is the difference between temporary progress and lasting development. This chapter begins with that recognition and argues that to understand reliability, we must look beyond the grid itself to the external conditions that shape its stability, its vulnerabilities, and ultimately, its role in supporting human dignity.

In the interconnected world of the 21st century, where technological advancements have reached remarkable heights, it's sobering to acknowledge that millions of people still live without electricity, which is for so many of us a fundamental, basic necessity. As dusk settles each day, and over each longitude of the globe, the dim reality unfolds in remote villages, sprawling urban slums, and marginalized communities in homes where the flicker of a light bulb remains a distant dream.

According to the latest estimate, approximately 759 million people worldwide—9% of the world's population [1]—lack access to electricity. The Rockefeller Foundation's "Electrifying Economies" initiative provides another stark metric: 1.5 billion people worldwide have *unreliable* access to electricity. This staggering number represents a significant challenge in achieving global development goals, such as those the UN SDG 7[1], which at the time of this writing, is not on track for attainment by 2030 [2]. Sub-Saharan Africa and parts of Asia, particularly rural areas, bear the brunt of this disparity: in sub-Saharan

[1] The United Nations' Sustainable Development Goal 7 (SDG 7) seeks to ensure access to affordable, reliable, sustainable, and modern energy for all by 2030. Meeting this development goal would also provide positive reinforcement and support to a number of the remaining 16 SDGs, including: 4 Quality Education, 6 Clean Water and Sanitation, 8 Decent Work and Economic Growth, 9 Industry, Innovation and Infrastructure, 10 Reduced Inequalities, 11 Sustainable Cities and Communities, and 16 Peace, Justice and Strong Institutions.

Africa alone, nearly 600 million people live without electricity, making it the region with the highest proportion of energy poverty.

For those living without electricity, daily life is marked by hardships that extend far beyond the absence of light after sunset. Cooking without modern energy sources such as electric stoves or reliable refrigeration, studying in the dim glow of kerosene lamps, and navigating health facilities without adequate power for medical equipment are everyday challenges. The lack of electricity also perpetuates cycles of poverty by hindering economic productivity and limiting access to education and information. The absence of electricity directly impacts healthcare services, jeopardizing the effectiveness of treatments and the storage of life-saving vaccines. Maternal and infant mortality rates are higher in areas without reliable electricity, where complications during childbirth and preventable diseases often claim lives that could have been saved with access to proper medical care. Energy also plays a significant role in providing access to clean water—an issue faced by at least 4.5 billion people [3]. Despite the magnitude of the challenge, concerted efforts by governments, international organizations, and non-governmental organizations (NGOs) are making strides towards bridging the energy gap with initiatives ranging from electrification projects powered by renewable energy sources such as solar and wind to policy reforms aimed at enhancing energy infrastructure and affordability.

This ambitious target of ending energy poverty and achieving universal access to electricity is not only a matter of infrastructure development but also requires addressing financial, political, and social barriers that hinder progress. The quest for universal access to electricity is an economic necessity and a moral imperative; but it demands collective action and innovative solutions to overcome entrenched challenges, particularly those related to ensuring reliable electricity supply in regions riddled with co-morbidities of an unstable or underdeveloped society. For one, we have largely failed thus far in articulating a commitment to reliable electricity (beyond SDG7) in regions experiencing conflict, natural disasters, economic collapse, corruption, or refugee crises to name a few. Bridging the energy divide and supplying reliable electricity remains fundamentally essential to fostering inclusive growth, empowering communities, and advancing human dignity across borders and generations to ensure that no one is left in the dark.

With this in mind, I would also like to dedicate this book to the nearly 800 million people on the planet who do not yet have access to electricity, and the millions more who have poor, unsafe, or intermittent electricity, during an age when it seems that everything we do can be wholly simplified and enhanced by power and light. I hope that in one way or another, this book helps to bring more power and light to the world.

1.2 The Role of Reliability

With the rapid transformations affecting the grid and worldwide acceleration of electrification, in addition to changing demand profiles for electricity (such as with record numbers of data centers coming online), exponential growth in adoption of technology relying on electricity (such as electric vehicles, laptops and smart phones), and continuing influx of renewables, the electric industry's understanding of reliability is shifting and expanding. Even so, the most cutting-edge and state-of-the-art analysis is leaving behind countries which have lagged in technological adoption, especially the underdeveloped countries (as defined by United Nations). Our most aggressive advancements at the forefront are not particularly serving the planners who wish to understand their own weakened or destabilized systems and also fail to inform our global development needs.

According to the annual SOR published by NERC the reliability of the *interconnected BPS* is concerned with the adequacy of resources and operating reliability. NERC is primarily concerned with ensuring reliable transmission to avoid a largescale collapse and has done a fantastic job at ensuring the North American electric grid's reliability through development of expertise, research, data, and large-scale analysis (much of which is openly available on NERC's website).

NERC's focus on bulk power system reliability necessarily places distribution-level and international reliability considerations outside its scope. It is not the case that reliable electricity only requires a reliable transmission grid; we know that distribution level outages can have the same impact on the end-user as a larger scale outage event. This is a sphere of knowledge outside of the jurisdiction of the ERO framework and thus which industry, academia, markets, policy institutions, think tanks, and others concerned with the provision of reliable electricity, including the readers of this book, must address.

The worldwide perspective on electrification and grid reliability paints a different picture of reliability than what we see in North America and is perhaps even more demonstrative of the need for increased attention to describing the reliability of diverse grid situations, such as that of a post-civil war country, one facing a severe and long-term economic downturn, or island states challenged by severe resource constraints. These electric grids and their planners, organizers, engineers, and operators face challenges very different from those faced by the highly developed countries of North America and Western Europe when it comes to grid reliability. Although operational reliability and resource adequacy have classically been the primary points of focus in the study of grid reliability, the introduction of external factors, such as extreme weather events, policy decisions and initiatives, crime and vandalism, socioeconomic conditions, and more, are having an increasingly robust and present impact on the reliable function of the electric grid. This is the primary focus of this book.

Throughout the text, discussions are mostly centered around external factors influencing reliability, and the path those factors follow to ultimately affect the internal conditions of the grid. The purpose of this approach is to illustrate the need to expand the purview of analysts, engineers, policymakers, and leaders in the energy sector, to take a broader approach to understanding and ultimately influencing grid reliability.

1.3 Understanding the Meaning of Reliability

A reliable electric grid is one that produces electricity for end consumers to utilize any time it is needed or desired. The inverse of reliability is failure. Lack of reliability implies the lack of availability of electricity, or lack of ability of end users to consume electricity when needed or desired. Historically, high-level discussions on grid reliability have been limited to operational reliability, with a primary focus on the requirements such as inertial stability and frequency stability. This is the primary arena for asset owners and operators, grid planners, and balancing authorities; the conditions and structures internal to the electric grid.

This perspective on grid reliability however has not provided enough breadth to describe some critical failure or widespread unreliability conditions; according to an article in *Energy Informatics* [4] a review of the state of the art of energy research shows that grid reliability is in fact an understudied research area.

Understanding a complex system is a major challenge; to make matters even more exciting, the electric grid is itself a complex system which is interconnected with many other complex systems—with dynamics that to some seem as intricate and mysterious as the movements of the ocean or the sudden development of turbulent flow in the smoke emanating from a candle **(Figure 1.1)**.

Figure 1.1 A Schlieren photograph showing the mysteriously organized transition from laminar to turbulent flow in the plume rising from a candle.
Image Licensor: Gary Settles, CC BY-SA 3.0 https://creativecommons.org/licenses/by-sa/3.0, via Wikimedia Commons. No changes were made to the original image.

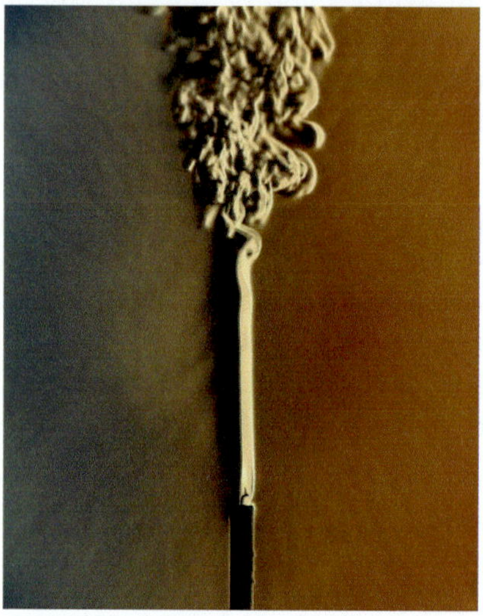

1.4 Challenges to Understanding Reliability

The foremost challenge of understanding the reliability of an electric grid is that each electric grid system is built uniquely, has unique dynamics, and faces distinct risk and threat environments, likely paired with anomalous dynamics of feedback and dampening cycles that are not seen in other systems.

For this reason, it is understandable that globally, research on grid reliability sometimes focuses heavily on the specific applications and data points which are relevant to a specific grid and is hyper-local in nature. Comparative analyses, which highlight the strengths and weaknesses of different grids and their respective environments, or generic and broad strokes on grid reliability, are extraordinarily limited. This is not because there is no interest in a comparative analysis—experts are stirring to understand grid reliability globally, but especially in a comparative sense as there is intrinsically great value in the ability to level-set and compare or benchmark against peers globally [5].

As an example, in the year of publication of this book, there is currently a war in Ukraine; and the electric grid there is faring reasonably well. However, it is likely that it is not performing as well as one of its neighbors, such as Belarus, who is not suffering bombardment, shelling, or other attacks on infrastructure. By the definitions that follow for reliability, Ukraine appears to have a less reliable electric grid, since it is not providing electricity to the end consumer as reliably as its neighbor, however the underlying dynamics of the situation are of paramount importance. In fact, the ability for a nation experiencing armed conflict to maintain any degree of electric reliability is an outstanding accomplishment. Being able to understand the currents of resilience, security, and operational reliability which are working together to provide stability in the form of electricity services would be a massive advancement, as it would allow planners across the globe to make a concentrated effort in enabling the same stability even under similarly complicated circumstances.

The following chapters will explore the influence of disaster-level events, the interactions between affordability and reliability, the influence of institutional design, and the relationships between reliability and state fragility, geopolitical and security risks, providing the reader with a broad foundational structure that can be used to enhance understanding of the complex dynamics of grid reliability.

1.5 Questions

(1) What does it mean for electricity to be a moral imperative?
Consider whether grid reliability should be a human rights issue in global development. Who bears the responsibility to ensure equitable access to reliable electricity?

(2) How does the meaning of "reliability" differ between a war-torn country like Ukraine and a highly developed grid like the U.S.?
In what ways should reliability metrics be adapted to different geopolitical contexts?

(3) NERC's focus on large-scale transmission reliability has been effective for North America. But what are the risks of applying this framework to underdeveloped regions?
Propose a modified or expanded reliability paradigm that could support local or distribution-level challenges in low-resource settings.

(4) In what ways is the electric grid an example of a complex adaptive system?
Identify feedback loops, interacting sub-systems, and unpredictable behaviors that distinguish electric grids from simpler infrastructure systems.

(5) The text argues that the electric grid is "interconnected with many other complex systems."
Choose one such system—healthcare, water, transportation, or internet—and analyze how disruptions in that system might affect grid reliability, and vice versa.

(6) Compare and contrast the goals of SDG7 (clean, affordable energy) with the concept of grid reliability.
Are they aligned? Where do they diverge? Can universal electrification occur without reliability, and should it?

(7) What barriers—technical, political, or institutional—prevent us from creating more comparative reliability frameworks across countries?
How could global cooperation improve our understanding of diverse reliability environments?

(8) How might your own understanding of "reliable electricity" be shaped by your geographic or economic context?
Reflect on how expectations for reliability differ between communities, nations, or user types.

(9) If you had to create a new global indicator for "grid reliability equity," what would it measure?
Draft a few possible components or sub-metrics and justify their inclusion.

(10) This book seeks to "broaden the lens of reliability beyond system dynamics to include external pressures.
What are the benefits and risks of this expanded view? Could too broad a definition dilute the engineering rigor of reliability studies?

References

[1] United Nations, "High-level political forum on sustainable development: Progress on the Sustainable Development Goals," [online], 2023.

[2] IEA, IRENA, UNSD, World Bank, WHO, "Tracking SDG 7: The Energy Progress Report," World Bank, Washington, DC, 2024.

[3] M. Walton, "Energy has a role to play in achieving universal access to clean water and sanitation," 2018. [Online]. Available: https://www.iea.org/commentaries/energy-has-a-role-to-play-in-achieving-universal-access-to-clean-water-and-sanitation.

[4] V. Sultan and B. Hilton, "Electric grid reliability research," *Energy Informatics,* 2019.

[5] P. Mauldin, "U.S. Power Reliability: Are We Kidding Ourselves?," T&D World, 2015. [Online]. Available: https://www.tdworld.com/grid-innovations/article/20966117/us-power-reliability-are-we-kidding-ourselves.

CHAPTER

2

Disaster Exposure, Affordability, and Institutional Design

Abstract

This chapter examines two foundational external spheres that shape electric grid reliability: disaster-level natural hazards and the economic and institutional conditions that govern affordability. As climate-driven events intensify, resilience must be understood not only as a technical attribute, but as a system's capacity to withstand and recover from shocks beyond operator control. At the same time, the chapter challenges a pervasive assumption in energy policy and industry discourse: that improving reliability necessarily reduces affordability. Using cross-national data comparing cost of electricity, affordability metrics, and quality of supply across more than one hundred countries, this chapter demonstrates that the relationship between reliability and affordability is not inherently inverse, and that a grid can be both unreliable and expensive. Instead, the analysis identifies underlying drivers such as macroeconomic stability, governance structures, and institutional capacity as stronger predictors of reliability outcomes. Together, these findings expand the reliability planning lens to include hazard exposure, recovery thresholds, cost structures, household electricity behaviors, and sector governance, establishing reliability as a socioeconomic and environmental outcome, not only as an engineering one.

DOI: 10.1201/9788743813491-2

2.1 Introduction

Reliability is often described as a technical condition of the grid, but it is equally an economic and environmental one. A system may be fully electrified yet still fail its population if power is unaffordable, intermittent, or fragile in the face of disturbance. Likewise, households do not experience reliability in kilowatts or inertia—they experience it in the rhythms of daily life: whether they can refrigerate food, study at night, heat or cool their homes, or rely on the grid when they need it most. This chapter begins by examining how disaster-level natural events reshape the stability of electric systems and then turns to the economic dynamics of cost and affordability, demonstrating that reliability is constrained or enabled by the financial conditions under which electricity is consumed. Reliability is therefore not only engineered; it is paid for, governed, and lived. Understanding reliability in this broader frame allows planners to identify the mechanisms through which resilience can be strengthened—without assuming that affordability and reliability must be in conflict.

2.2 Climate Change and Resilience to Disaster-level Events

As disaster-level events become more frequent and severe, the ability of the electric grid to preserve its function and operation during disturbance events is critical to its survival. Findings in support of the growing threat of disaster-level natural events [1] coherently cite the primary source of service disruptions in the United States: "between 2003 and 2012, 80% of all outages were caused by weather." Some frameworks for reliability implicitly consider resilience as a primary factor, however, resilience is a quality that depends on the features of the major events that require resilient responses, meaning that resilience is an adaptation to external inputs. For a built system which requires significant time, investment, and planning, and is largely static over short time-horizons, it is extremely important to develop a deep understanding of the external events to which the system must instantaneously adjust. Rather than excluding major events from reliability criteria, planners must continue to expand their focus on actively studying and preparing for extreme events which occur with increasing frequency.

Heat waves, cold snaps, wildfires, flash floods, glaciers and permafrost thaw, tropical cyclones, and storm systems will continue their trajectory of increasing frequency of severity. Not only will disaster-level storms continue to wreak havoc on our electricity system, but record heat waves and cold snaps will continue to push the envelope in terms of cooling and heating needs, requiring more electricity each year as record temperatures reach new extremes.

These changing norms have brough multi-dimensional and cross-sector impacts across large regions within North America and will continue to threaten development goals across the globe.

According to Sunny Wescott, Chief Meteorologist at DHS CISA, natural disasters pose a particular threat to communications infrastructure which impacts other sectors such as financial services, the IT sector, emergency response services, transportation sectors and energy/electricity and negative impacts will continually be exacerbated by the supply chain issues that ensue following a disaster-level event. How will our systems fare given that natural disasters across the globe are increasing in frequency and intensity and are projected to continue increasing in the coming decades?

2.2.1 Key examples

2.2.1.1 Island-wide blackout in Puerto Rico, 2025

Most of the island of Puerto Rico, including over a million homes, businesses, and critical infrastructure facilities (such as water/wastewater facilities and hospitals) were without power on New Year's Eve and New Year's Day in 2024/2025. According to the director of the Puerto Rico Power Authority, the failure of a single power cable resulted in the cascading blackout across the island. Experts have attributed the ongoing fragility of the electric grid in Puerto Rico to the devastation from two quick-succession high-intensity hurricanes in 2017 (Hurricane Irma, Category 5 followed by Hurricane Maria, Category 4). Of course, Puerto Rico's progress has been further hindered since 2017 by additional storms, including smaller-level tropical storms and hurricanes, including Hurricane Fiona in 2022. According to earth and space science researchers, hurricanes in the Atlantic have been increasing in intensity since the 1980s, and many analysts attribute this to increasing ocean surface temperatures [2].

Puerto Rico has struggled to adequately rebuild since Hurricane Maria and faces the unfortunate challenge in the coming decades of increasingly intense and frequent hurricanes. Rebuilding efforts in Puerto Rico have faced further challenges, including lack of institutional capacity for managing disaster recovery programs, according to the GAO [3].

2.2.1.2 2021 historic heat wave and persistent drought in Pacific Northwest

In June 2021 an unprecedented heatwave (see Figure 2.1) in the Pacific Northwest, characterized by temperatures of 116–118 °F in the major metropolitan area of Portland, resulted in devastating wildfires and record temperatures across the region, including the record national temperature for Canada of 121 °F—the highest temperature ever recorded at the 45 °F latitude [4].

This "heat dome" resulted in $8.9 billion of damage, primarily to crops and vegetation, and 229 attributable deaths. Power outages to transportation infrastructure resulted in outages to the public transportation system [5].

Figure 2.1 2021 Pacific Northwest heatwave temperature visualization by National Oceanic and Atmospheric Administration. Image source: NOAA climate.gov

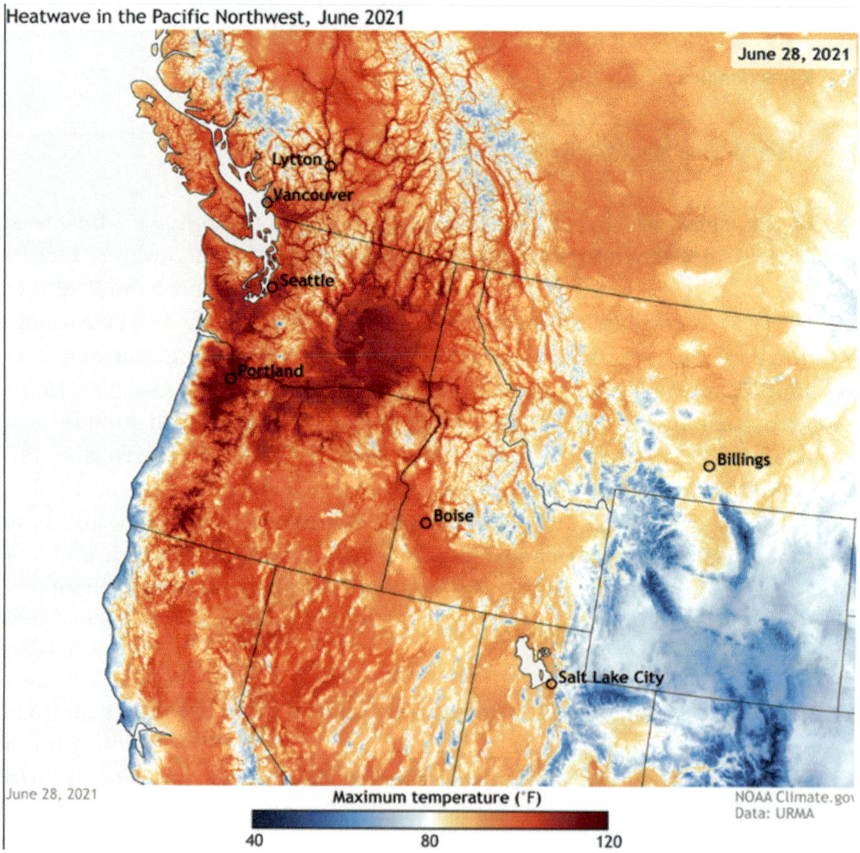

Furthermore, the historic drought that accompanied the heat dome resulted in reduced water supply and lower hydropower generation in California, with hydropower generation making up 48% less than the previous 10-year average for the region. Similarly in the Pacific Northwest, hydropower generation was 14% lower than the 10-year average for the state—demonstrating the direct risk to electric reliability posed by disaster-level events [6].

Swiss Re reports that extreme temperatures, likely to increase in the coming decades, can also degrade performance of installed infrastructure, including large power transformers, and underground distribution infrastructure, and decrease the lifetime of these components by 20–40% [7].

As such, reliability planners must consider the risks of disaster-level events and implement mitigations that leverage large-scale resilience such as regional electricity interchange, and direct consideration of fuel availability (water for hydropower generation) in order to continue to meet demand under increasingly extreme circumstances.

2.2.2 Metrics for quantifying

In the early 2000s, weather-related research concerning electric utilities in North America reached a turning point—the grid itself in the United States was in transition, and the face of grid reliability planning as we see it today was nascent. During those years, industry analysts found that the inclusion of outages related to weather events were adding high levels of variability to utility performance data. When research confirmed correlations between weather-related data points and reliability indices, the persisting solution was to normalize the weather-related data to essentially lighten its impact on reliability indices. This reveals the core thinking that remains intact today with respect to weather related events: outages are inevitable; they're a function of bad weather, which is simply the luck of the draw, and cannot be reasonably avoided.

However today, as weather, climate impacts, and other disaster-level natural events increasingly impact the electric grid, reversing this thinking becomes critically important. This section quickly outlines a metric that can be used to measure rather than ignore or normalize the impact of weather events on grid reliability.

2.2.2.1 Caveated use of CAIDI, SAIDI, SAIFI, etc.

Where the common practice is currently to use either the pure reliability index, or the reliability index with major event days excluded, the analyst should arrive at the practice of instead always producing both values. However, these indices are still limited to providing a reflection of performance, not a reflection of the risk that the system faces.

2.2.2.1.1 Annualized frequency

It is critical to understand the actual exposure to disaster-level events and hazards. The most usable composition for use within a general reliability framework is the annualized frequency of hazards/disaster-level events, or the annual probability of exposure. The U.S. Federal Emergency Management Administration (FEMA) developed a fantastically comprehensive risk ledger for every county in the United States, providing a risk rating index across over a dozen types of natural hazards. Reading the technical documentation [8] for FEMA's National Risk Index is a major lesson in understanding the risk in a broad operating area, and an extremely useful resource for planners, analysts, or policymakers in a region with less of a centralized risk planning function. Within this framework by FEMA is a golden nugget—the Annualized Frequency. The simplest version of this value provides an annual probability for a given hazard, and is given by:

$$p = \frac{\#\,of\ recorded\ occurences}{period\ of\ record}\left[number\ per\ year\right] \tag{2.1}$$

The UN's Sendai Framework for Disaster Risk identifies three key elements: exposure to hazards, vulnerability and capacity, and hazard's characteristics [9]. Integrating just one of these key elements, exposure to hazards, into a reliability outlook would involve the use of the aforementioned probability—the annual probability for a given hazard. Expanding on this approach considers both the hazard's characteristics and the vulnerability and capacity of the system by assessing the system's capacity for recovery for a particular hazard type. For example, the subsystem in Candyland can withstand 48 days of severe drought [including severe drought threshold/definition] before experiencing unacceptable impacts to reliability [including detailed criteria for how this hazard meets the unacceptable level].

Hazard characteristics: severe drought—i.e. meets D1 Drought Classification as defined by National Drought Mitigation Center [10].

Vulnerability: unacceptable impacts to reliability

Capacity: 48 days @ D1

This formulation establishes a current capacity baseline and a hazard threshold, which narrows down the subset of events within the hazard domain that will be counted as an "exposure". Then, the annualized probability can be adapted by assessing the number of recorded occurrences which have historically exceeded the capacity of the system in the same operating area or a similar operating area. Finally, the collection of exposures for each hazard type given the system capacity can in turn be integrated into a Bayesian Network or combined with probabilistic approaches—Markov processes, Monte Carlo simulations—in order to provide additional inputs to an overall reliability analysis.

The hazard domains are according to the National Risk Index are:

(1) Avalanche
(2) Coastal flooding
(3) Cold wave
(4) Drought
(5) Earthquake
(6) Hail
(7) Heat wave
(8) Hurricane
(9) Ice storm
(10) Landslide
(11) Lightning
(12) Riverine flooding
(13) Strong wind
(14) Tornado
(15) Tsunami
(16) Volcanic activity
(17) Wildfire
(18) Winter weather

For each hazard domain, determine which ones apply to the area or region of study for the electric grid. Then, establish within each hazard domain a maximum threshold characteristic that the system can withstand. For example, in a coastal area, determine the maximum amount of coastal flooding (in ft or m) that can be withstood by the system without significant reliability impacts. The collection of hazard domains and their respective characteristics and thresholds can be used to determine the exposure to each of these hazards by quantifying the likelihood of exposure to events meeting the specified level of hazard.

2.2.3 Conclusion

The approach described herein allows reliability planners to integrate the potential impacts of disaster-level events into their assessment and analysis processes. As these events become increasingly frequent and impactful, an expanding focus on impactful features within the reliability landscape will enhance the ability to plan and prepare for such events.

2.3 Affordability and Reliability

For those who have not experienced it first hand, it is a difficult condition to comprehend—coordinating your daily activities around the ability to use electricity at some times of the day but not others, unexpected loss of the ability to carry out functions we take for granted, such as listening to the news, using an overhead lamp to read a book in the evening, wash clothes with an electric washer, or simply being able to afford to keep your home at a comfortable temperature. Chapter 1 summarizes the current state of electricity access across the globe, but makes the salient point that many with access to electricity still don't experience *reliable* electricity, and even IEA's definition of "access"—the ability to consume at least 1250 kWh per year [11]—is only around 10% of the average household consumption in the United States [12]. I have a sneaking suspicion that when affordability is thrown into the mix, the number of people experiencing unreliable and/or unaffordable electricity service grows exponentially. Thus affordability is not a characteristic of reliability that is anticipated in most of the Western democracies, where social services, non-profits, local charitable funds, or help from community members, is likely to lift those experiencing extreme poverty into more livable situations, except in the most extreme cases. In other parts of the world, however, what we consider to be extreme poverty is the norm or would even be considered a middle-class living situation. In much of South America for example (which I speak to in a personal sense from firsthand anecdotes rather than as a researcher) it is rare to find a home that runs a climate control system year-round. The cost of electricity relative to the median household income is unattainably high, and thus electricity is reserved for necessities such as lighting and the refrigeration of perishable goods. In most homes, a clothes dryer is reserved for use only when a quick turnaround is needed, otherwise the clothes would be dried on a clothesline. A dishwasher is a luxury that only the elite might possess, but in those cases, it would still be more likely for those households to hire a live-in maid as it would be similar in cost to running a dishwasher every time the dishes are used and would save the time of loading and unloading.

With this picture in mind, I pose to the reader the following rhetorical question (rhetorical, because I hope that I have already implied its answer): is a system which is so expensive that its use is highly constrained a reliable and sustainable electric grid? What about a system constrained by crippling inefficiencies or poor institutional governance? Or a system exposed to numerous disaster-level events in a decade, or in a year? Or one burdened by a disproportionately demanding operating environment, such as violent fluctuations in inflation (multiplied by something like the order of hundreds or even thousands of what was experienced post-COVID in North America)? When an entire nation is experiencing the *force majeure* of a civil war, or other armed conflict, are those excluded and the system deemed reliable to a suitable extent?

When considering such situations, a distinct world emerges.

2.3.1 The fallacy of reliability impeding affordability

The position that affordability is a primary balancing factor with reliability is rampant, not just within the literature [13], but also as a common theme within business discussions in the risk management and reliability engineering spaces. It is a position affirmed by logic: it costs money to make enhancements, to invest in infrastructure, and to make systematic improvements in reliability, especially when those investments are redundancy measures. This position assumes that the costs will be passed on to the end user, resulting in higher electricity tariffs; it also assumes that the spending is effective at enhancing reliability. Although this may be true on a microeconomic level, such as individual utilities making extensive reliability, resilience, and/or redundancy measures and then increasing rates—the data available does not support this position at the largest scale.

If reliability and affordability were strongly linked, then countries with the most reliable, high quality electricity supply would also be the most expensive. This section explores the data available to discuss the presence or absence of relationships between cost of electricity, or affordability of electricity, and quality of electricity. The development of an “affordability” metric enhances the analytical landscape in this space, allowing a careful look at the driving forces and conditional mechanisms contributing to these dynamics. Finally, the question is answered: why is it imperative to the survival of the system itself to ensure a reliable and affordable electricity supply?

2.3.2 Comparison of cost of electricity to quality of electricity

Based on a sample of 133 different countries for which both quality of electricity supply[1] and cost of electricity to domestic users[2] was readily available, statistical analysis of the data, including linear, second and third order polynomial, exponential, and logarithmic regressions, and t-tests on the regression with the best R^2 value, all confirmed that it is unlikely that a strong relationship exists between quality of electricity supply and cost of electricity on the scale being studied.

In this case, the null hypothesis states that there is no relationship between cost of electricity and quality of electricity.[3] Here, these two metrics (*cost of electricity* and *quality of electricity*) are stand-ins for affordability and reliability. This analysis relies on many assumptions—the first of which being that these two are suitable stand-ins, which is certainly an arguable point. Because of income inequalities that exist within the countries in the dataset and between the countries included, a comparison of USD/kWh does little to illuminate how affordable electricity will be to the population or particular subsets therein. This assumption is addressed in the following section where an affordability metric is developed and studied.

Despite this gap, an analysis of the presence or absence of a relationship between cost and quality of electricity still illuminates the patterns within the dataset and paints an interesting picture.

[1] Quality of electricity supply data is available from the World Economic Forum for 134 different countries. It is a single value varying from 1 to 7 and is based on survey results of the population in each country, where 1 is extremely unreliable and 7 is extremely reliable.

[2] Cost of electricity in USD/kWh was compiled from over a dozen sources, using the most recent available data accessible from open sources on the internet. All but 14 of the countries' cost of electricity data was collected from a single source (Cost of Electricity by Country 2025. World Population Review. Retrieved August 20, 2025, from https://worldpopulationreview.com/country-rankings/cost-of-electricity-by-country), and the remaining 14 were collected from a variety of sources.

[3] Null hypothesis H_0: There is no relationship between quality of electricity supply and cost of electricity for domestic users. Alternative phrasing: the cost of electricity in a given country has no effect on the quality of electricity supply for domestic users in that country.

2.3.2.1 Cost of electricity data characteristics

Approximately 90% of the countries sampled lie within the range:

$$\$0.00 < cost < \$0.35$$

and 50% of the countries lie within the range:

$$\$0.00 < cost < \$0.15$$

There is no upper bound to the data discoverable within this set, although it is probable that many of these economies are experiencing the diminishing marginal utility point guiding the discretionary choices made by consumers to utilize less electricity for non-critical purposes such as air conditioning and running appliances. The statistical values below summarize the data plotted on a scatterplot in Figure 2.2.:

Lower bound:	$0.00 (exclusive)
Minimum:	$0.002 (Iran)
Maximum:	$0.980 (Yemen)
Mean:	$0.168 ± $0.022

Figure 2.2 Quality of electricity supply plotted against cost of electricity in U.S. Dollars for 133 different countries around the world. Data combined from a variety of sources.

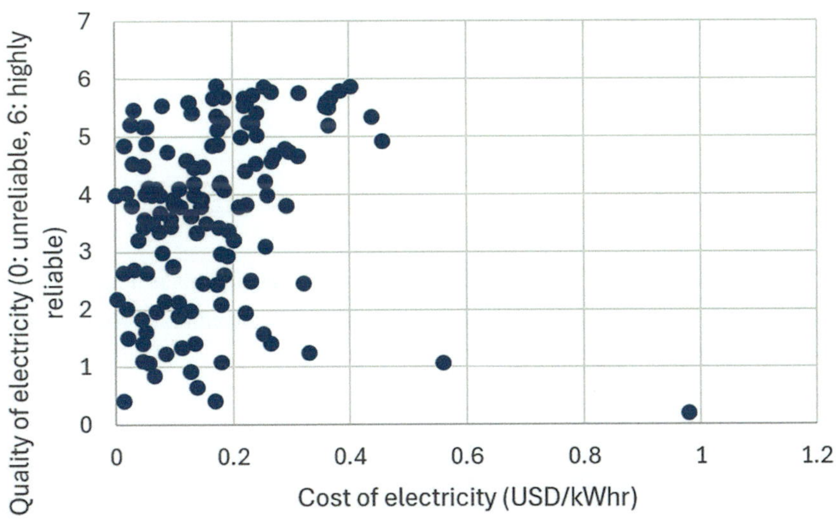

2.3.2.2 Regression results

As summarized in Table 2.1. Regressions were conducted on the data with cost on the x-axis as the independent variable (cost-driven model), then reversed with cost on the y-axis as the dependent variable (quality-driven model). Regression was also conducted with a revised dataset following the removal of outliers with a z-score higher than ±3.0, which included just two data points:

- Benin, $0.56/kWh, z = 3.020073
- Yemen, $0.98/kWh, z = 6.259531.

With exception of the logarithmic regression, the quality-driven models outperformed the cost-driven models, and the removal of the >3.0 *z*-score outliers also improved the R^2 values across the board. Calculating the *p*-values for the best performing R^2 value (the third order polynomial regression on the dataset excluding outliers) produced insignificant *p*-values *($p>0.05$)*, meaning that this test does not provide evidence significant enough to justify rejecting the null hypothesis for this relational model. Further testing on the second-best performing model (the second order polynomial regression on the dataset excluding outliers) produced mixed results, with significant *p*-values for the intercept and x^2 coefficients and an insignificant *p*-value for the x coefficient. Additionally, there is no underlying logical basis for a well-performing second- or third-order polynomials within a mostly economic problem with natural limits for both axes that are poorly represented by polynomials. Extreme values, such as outliers, changed the concavity of polynomial regressed for the cost-driven relationship, with a negative concavity regressed for the model with outliers and a positive concavity for the same model without outliers. Given this mix of outcomes, these tests are inconclusive, and the null hypothesis cannot be rejected.

Table 2.1 R^2 results of a variety of regressions for cost-driven and quality-driven relational models.

	Cost-driven model		Quality-driven model	
Type of regression	**With outliers**	**Without outliers**	**With outliers**	**Without outliers**
Linear	0.0183	0.1396	0.0183	0.1396
Exponential	0.022	0.1417	0.0299	0.1626
Logarithmic	0.0486	0.092	0.0022	0.0938
Second-order polynomial	0.1409	0.142	0.1688	0.193
Third-order polynomial	0.1445	0.1424	0.1844	0.2085

2.3.3 Household electricity burden: building an affordability metric

In order to more effectively quantify and compare the affordability of electricity in different countries around the world, this section defines a metric termed "household electricity burden" which describes the ratio between minimum household electricity use and the local GDP/capita. The metric is developed as follows:

(1) Estimate a minimum standard for monthly electricity usage for households in kWh (let A = annual household electricity usage).
(2) For each country, calculate the annual cost of electricity given minimum usage (step 1) and unit cost of electricity (let u = unit cost of electricity; and c = annual cost of electricity given A and u).
(3) For each country, calculate the ratio between annual cost of electricity (step 2) and GDP per capita (household electricity burden) (let y' = annual GDP/capita and $\mathcal{E}$ = household electricity burden)

Then, given A, u_i, and y', the household electricity burden in country i is given as

$$c_i = A * u_i$$
$$\varepsilon_i = \frac{c_i}{y'} = \frac{A * u_i}{y'} \qquad (2.2)$$

This produces a reasonable, yet imperfect comparative metric which normalizes equitable accessibility of electricity across countries based on local economic conditions. While it does improve the approach to addressing income inequality and energy inequity, particularly between countries, these inequities within a country are still not addressed. The use of GDP per capita provides insight into the country's overall economic health rather than the individual household's, which would be better represented by median household income. However, data collection limitations with respect to median household income made the use of GDP per capita a more feasible data point.

2.3.3.1 Estimating a minimum standard for household electricity usage

In order to determine and compare the affordability of electricity across different countries, where both household incomes and cost of electricity per kWh vary, a process is suggested for determining a minimum electricity usage standard. The International Energy Agency tracks access to electricity across the world, with the minimum threshold for considered as "having access" to electricity being usage of 1250 kWh per year. Table 2.2 presents an alternative minimum electricity usage standard which is driven by quality of life improvements.

Table 2.2 Estimate of a minimum quality of life standard usage of electricity beyond the 1,250 kWh "access" standard by the United Nations

Appliance[4]	Monthly usage estimate
Internet router	7.3 kWh
Refrigerator	45 kWh
Electric cooking appliance, 2 hours per day	91.25 kWh
Washer/dryer, 4 loads per week	101.48 kWh
Fan, 8 hours per day	7.3 kWh
Lighting, 5 hours per day	4.5 kWh
Subtotal	256.83 kWh/month
Total	3081.96 kWh/year

For this comparison, there were more data points available for GDP/capita than for the quality of electricity which was analyzed in the previous section. Some characteristics of the initial sample, which includes 200 countries, are summarized in Table 2.3.

Table 2.3 Characteristics of unit cost of electricity and household electricity burden data sets.

	Data set characteristic	
Data characteristic	USD/kWh	Household electricity burden (annual cost/ GDP per capita)
Mean	0.196333	0.282252
Standard error	0.010679	0.060527
Median	0.1735	0.048403
Mode	0.33	#N/A
Standard deviation	0.151023	0.855985
Sample variance	0.022808	0.73271
Kurtosis	4.002547	76.7907

[4] Appliance usage estimates are based on data available from Silicon Valley Power [32].

Data characteristic	Data set characteristic	
	USD/kWh	Household electricity burden (annual cost/ GDP per capita)
Skewness	1.472808	7.831829
Range	0.978	9.68477
Minimum	0.002	0.001795
Maximum	0.98	9.686564

The minimum and maximum values remained unchanged, and average values increased slightly given the addition of 70 data points. Using the "unaffordability" metric as a basis for calculating the *z*-scores, four outliers with $z > 3.0$ were identified (Table 2.4).

Table 2.4 Outliers within household electricity burden dataset identified by *z*-score greater than 3.

Country	Unit cost of electricity (USD/kWh)	Household electricity burden	*z*-score
Somalia	0.5	336.08%	3.606
Burundi	0.133	369.94%	4.002
South Sudan	0.336	409.84%	4.469
Yemen	0.98	968.66%	11.014

2.3.4 Comparison of household electricity burden and the quality of electricity

This section returns to the quality of electricity data (and thus again restricts the dataset to the 132 countries for which data is available), conducting a variety of regressions to understand the relationship or lack of relationship between the GDP-driven "household electricity burden" metric and the quality of electricity.

Note that the household electricity burden describes the estimated minimum share of per capita GDP spent on electricity given a minimum consumption value as estimated in the section above. Thus the higher the value, the less affordable electricity is estimated to be for the average household in that country.

In this case, the null hypothesis states that there is no relationship between the affordability metric and the nation's quality of electricity supply. The cost-quality (and quality-cost) regressions in the previous section posited a light positive relationship between the cost of electricity and reliability of supply (termed quality of supply), but tests were generally inconclusive.

The affordability-quality (and quality-affordability) regressions are summarized in Table 2.5.

Table 2.5 Performance (R^2 results) of a variety of regressions for affordability-driven and quality-driven relational models between quality of electricity supply and household electricity burden.

	Affordability-driven model		Quality-driven model	
Type of regression	**With outliers**	**Without outliers**	**With outliers**	**Without outliers**
Linear	0.1554	0.294	0.2554	0.294
Exponential	0.3029	0.3582	0.2886	0.2515
Logarithmic	0.4955	0.4521	0.3086	0.2624
Second-order polynomial	0.3057	0.4085	0.2487	0.3027
Third-order polynomial	0.396	0.4753	0.3177	0.3347

These models performed better than the cost-quality regressions and, most notably, show a strong opposite directionality: with 12 of the 20 relationships tested having an inverse relationship between increasing share of spending on electricity and quality[5] of electricity (linear, exponential, and logarithmic regressions for affordability- and quality-driven models both with and without outliers). In other words, the regressions suggest that nations with higher-quality electricity supply may tend to have low cost of electricity burdens relative to national GDP per capita.

The highest performing regression was the affordability-driven model with outliers, meaning that extreme values were well-represented by the model, unlike the cost-quality models. An inverse linear, exponential, or logarithmic model is well-suited to the underlying relationship, whereas there is no logical basis for high performance of second- or third-order polynomials, suggesting that the high-performing second and third order polynomials without outliers may be overfitted. For the highest performing models, derivative tests[6] show a

[5] The second- and third-order polynomial regressions had mix-signed coefficients but positive concavity for all x-values within the sample range.

[6] First derivative testing reveals positive concavity with a local minimum within the x-range of sample points for second order polynomials. Second derivative testing reveals inflection points within the x-range of sample points for third order polynomials.

directionality change within the x-range of sample values. This is inconsistent with the underlying logical basis given the variables being studied.

The best performing regression yielded the following logarithmic relationship between the affordability metric, x, and the quality of electricity supply, $y(x)$ of the following form

$$y(x) = c_1 \ln(x) + c_2$$
$$C_1 = -0.697$$
$$C_2 = 1.4969$$

giving

$$y(x) = -0.697 \ln(x) + 1.4969$$

Testing the p-value for this regression yielded results well below the significance level ($\alpha = 0.05$):

$$C_1 : 3.40321 \times 10^{-21}$$
$$C_2 : 1.32018 \times 10^{-10}$$

This can be interpreted as there being an extremely low probability that this sample occurred by chance and is strong evidence against the null hypothesis—that there is no relationship between quality of electricity and affordability of electricity. Additionally, given that $C_1 < 0$, there is a very high likelihood that a negative logarithmic relationship exists between affordability and quality of electricity.

2.3.5 Assessing the influence of GDP

Because of the strong influence of GDP on the affordability metric, the final test of this chapter examines the degree of influence that GDP has on the quality of electricity supply.

Plotting the quality of electricity supply on the y-axis and GDP per capita on the x-axis reveals a strong relationship between the two variables (Figure 2.3).

Conducting the same regressions as was done for previous variables, the $\ln(x)$ function performed uniquely well, with an R^2 value of 0.7448. Further analysis would be needed to find a best-fit model; although for this application, the fit was sufficient (with sufficiently small p-values) to reject the null hypothesis and conclude that it is very unlikely that the relationship between quality of electricity supply and GDP are random.

Figure 2.3 Quality of electricity supply plotted against GDP for countries around the world, indicating logarithmic relationship and moderately high R^2 value. Data from a variety of sources.

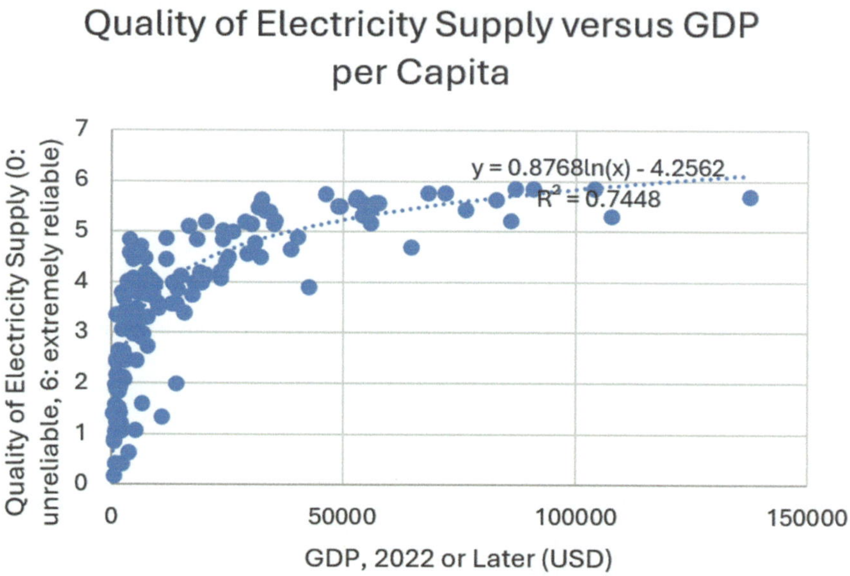

2.3.6 Conclusions: Discussion on driving forces and conditional mechanisms

Moving from cost of electricity, to the household electricity burden, to the GDP in relation to the quality of electricity supply produced increasingly coherent relationships. This chapter shows why the position that reliability and affordability are inversely proportional should be heavily questioned—instead planners should seek to understand the hidden structures driving quality of electricity supply. The primary tension between an electric grid, which must market its services well enough to receive payment in exchange, and the households making decisions regarding their electricity usage, is cost and affordability of those services. Low affordability of electricity leads to curtailed usage and thus less revenue, providing less capital for investment to enhance reliability or expand services. High affordability on the other hand strains usage, pushing the grid to its limits, which drives investment out of necessity. Rather than mutually competing fronts, reliability and affordability of electricity are likely both affected heavily by economic conditions both in the macroeconomic and microeconomic senses, with household decisions being tied heavily to the economic conditions and outlook for financial dynamics.

The ability for users to pay for the electricity they consume, and consume a formidable amount of it, is critical to the system's survival; it's also central to well-being and quality of life across the globe.

2.4 Undercurrent: The Challenge of Institutional Design for Energy and Electricity Sectors

2.4.1 A discussion on energy policy

The electric grid is an increasingly interdependent system-of-systems (SoS), heavily influenced by the conditions and elements outside of itself. In North America, at the forefront of the external conditions influencing the conditions of the grid is *energy policy*. This factor was newly identified in 2023 by the North American Electric Reliability Corporation (NERC) as a top risk to electric grid reliability [14].

The energy policy of today will impart powerful impressions on the power and electricity landscape in the coming decades [15]; and these impacts will be intensified by increasing growth in demand and the global geopolitical landscapes, which affect economic wellbeing and stability. Energy policy directly affects energy availability, and the reverberations from policy changes can have feedback effects on societal stability. Creative implementation strategies are needed to balance the numerous, often conflicting objectives of policymakers, while continuing to enhance the quality of life of the population served by the electric grid. Social, economic, environmental, political, and technological factors have been consistently presumed to be at odds with one another within the context of reliability and energy security [16] [17] [18]. The future of reliability analysis should implicitly balance the numerous conflicting demands in pursuit of reliability.

These hidden structures and mechanisms account for a significant portion of the risk faced by electric grids worldwide. Understanding these structures requires an expansion of the boundaries of the system to encompass the key external factors influencing grid reliability, and their relationships to the core functions of the grid. Furthermore, these multi-layered dynamics necessitate the development of a new prevailing theory to understand the mechanisms and structures at work—this is critical to the health of electric grids in the years to come.

One of the factors highlighted in this book is the need for neutral, reliability-focused oversight. There are so many competing priorities that reliability can be balanced with, including political agendas, economic needs, regional development priorities, environmental regulations, and more. The electric grid is a complex socio-technical structure that is influenced by these external pressures and forces, and requires a balanced institutional design focused solely on the well-being of the grid; in North America, NERC serves as the electric reliability organization (ERO) designated to develop and enforce reliability standards for the bulk power system under FERC oversight.

2.4.2 Understanding the Impacts of Institutional design

Reliability is a shared, governed, designed, engineered, constructed and maintained outcome and does not happen by chance. There are many examples today of centralized and fully verticalized public services, including electric services, controlled by the government at a national level. These even exist in smaller scales in the North American system.

In some fully centralized, large-scale public service models, electricity policy carries additional social or economic objectives which in some cases (i.e. when implemented responsibly) can be highly beneficial to stakeholders and consumers, as there are oftentimes layers of benefits beyond the public service being provided. However, when multiple objectives compete without a clearly prioritized reliability mandate, system performance can take a backseat and become strained. This dynamic reflects differences in institutional design and highlights the importance of clarity in mission setting for reliability focused governance. An imbalanced dynamic can emerge when institutional roles and responsibilities are not clearly defined or aligned. Ensuring reliable system performance therefore depends on clarity of authority and coordination across relevant institutions. This composition can occur any time there is an imbalance in the governance structure either from over-regulation or lack of governance. Experience across many systems suggests that clearly defined reliability responsibilities and sustained institutional capacity support long-term system performance.

2.4.2.1 Primary risk domains for grid governance

The primary risk domains for grid governance are:

- Clarity of authority—an intentional and clearly articulated structure of authority taking into account robust distributed authority when appropriate.
- Regulatory capacity—effective and appropriate oversight and guidance is provided through adequately staffed, highly trained, and well-funded regulatory bodies that are attuned to the

needs of the industry and expertly walk the balance between regulatory needs and enablement of investment.

- Market-policy alignment—policies support clear market signals geared towards industry actors (generators, retailers, investors, etc.) that result in investment decisions that enhance reliability.
- Decision coherence—harmony across jurisdictions is a force multiplier for reliability, with policies remaining stable over long time horizons, and agreement between levels of governance, resulting in adequate procedural coordination and alignment.
- Planning horizon—a strong planning horizon demonstrates a proactive approach to risk mitigation which integrates analysis of systemic conditions and strategic futures, reaching far beyond the short-term needs such as compliance, budgets, shock reactivity, and today's demand/supply balance.

2.4.3 Conclusion

Although not a primary factor, governance structures are a challenge closely related to the reliability of the electric grid. The coming chapters will discuss the factors which present more immediate concerns to the sector within the focus of reliability planning.

2.5 Exercises

(1) Based on the information provided in the affordability section, reflect on the potential biases present in the datasets. Discuss how these biases may impact an analysis that relies on these data sources.

(2) What types of decisions require the longest planning horizons in grid reliability?
How well are current utility and policy planning practices suited to anticipate these long-horizon risks? Propose improvements to institutional structures that would support more robust long-term planning.

(3) In what ways does fragmented governance increase systemic risk for the electric grid?
Identify a real-world example of a blackout or major disruption caused or worsened by poor decision coherence. How could governance reforms have prevented it?

(4) Evaluate the reliability consequences of rapid decarbonization.
Under what conditions can aggressive clean energy targets unintentionally reduce reliability? How can governance structures ensure that reliability and decarbonization efforts are mutually reinforcing?

(5) Propose a planning framework that integrates resilience to disaster-level events into utility resource adequacy assessments.
What metrics would you include, and how would this change how utilities invest in generation or infrastructure?

(6) Compare two grid planning institutions. How do their governance models differ in terms of decision coherence, planning horizon, and stakeholder accountability? Which model is more adaptable to future stressors?

(7) How can electric grid governance incorporate feedback from non-energy sectors (e.g., agriculture, telecommunications, emergency services)?
What governance models support cross-sector resilience planning, and how might they be applied to the electric sector?

(8) Design a "resilience stress test" for a regional grid system.
What scenarios would you model? What thresholds would define failure? How could results influence regulatory requirements?

(9) Critically assess the assumption that markets alone can ensure grid reliability.
Under what conditions does this hold true, and when does it break down? Provide examples of market-policy misalignment and propose governance responses.

(10) If you were advising a state public utility commission what institutional reforms would you recommend to strengthen their oversight of reliability?
Consider planning mandates, resilience integration, interagency coordination, and policy stability.

(11) Reflect on the ethical dimension of grid reliability planning.
Who bears the risk when planning horizons are short or misaligned? How should decision-makers weigh short-term cost increases against long-term systemic risk?

References

[1] V. Sultan and B. Hilton, "Electric grid reliability research," *Energy Informatics*, 2019.

[2] Knapp, Kossin, Vimont, Murnane and Harper, "A globally consistent reanalysis of hurricane variability and trends," Geophys. Res. Lett., vol. L04815, p. 34.

[3] Government Accountability Office, "Hurricane Recovery Can Take Years—But For Puerto Rico, 5 Years Show Its Unique Challenges," 14 November 2022. [Online]. Available: https://www.gao.gov/blog/hurricane-recovery-can-take-years-puerto-rico-5-years-show-its-unique-challenges. [Accessed 11 November 2025].

[4] R. R. White, S. Anderson, J. F. Booth, G. Braich, C. Draeger, C. Fei, C. D. G. Harley, S. B. Henderson, M. Jakob, C.-A. Lau, L. M. Admasu, V. Narinesingh, C. Rodell, E. Roocroft, K. R. Weinberger and G. West, "The unprecedented Pacific Northwest heatwave of June 2021," Nature Communications, vol. 14, no. 727, 2023.

[5] A. B. Smith, "2021 U.S. billion-dollar weather and climate disasters in historical context," NOAA, 24 January 2022. [Online]. Available: https://www.climate.gov/news-features/blogs/beyond-data/2021-us-billion-dollar-weather-and-climate-disasters-historical.

[6] U.S. Energy Information Administration, "Today in Energy: Drough effects on hydroelectricity generation in western U.S. differed by region in 2021," 30 March 2022. [Online]. Available: https://www.eia.gov/todayinenergy/detail.php?id=51839#:~:text=In%202021%2C%20a%20historic%20drought,the%20Northwestern%20River%20Forecast%20Center.

[7] M. Chatterjee, "Avoiding gridlock: The Impact of climate on electric grids," Swiss Re Institute, 27 June 2024. [Online]. Available: https://www.swissre.com/institute/research/topics-and-risk-dialogues/climate-and-natural-catastrophe-risk/impact-of-climate-on-electric-grids.html.

[8] C. Zuzak, E. Goodenough, C. Stanton, C. Mowrer, A. Sheehan, B. Roberts, P. McGuire and J. Rozelle, "National Risk Index Technical Documentation," Federal Emergency Management Agency, Washington, DC.

[9] United Nations Office for Disaster Risk Reduction, "What is the Sendai Framework for Disaster Risk Reduction?," United Nations, [Online]. Available: https://www.undrr.org/implementing-sendai-framework/what-sendai-framework#targets.

[10] National Drought Mitigation Center at University of Nebrasks Lincoln, "Drought Classification," University of Nebraska-Lincoln, [Online]. Available: https://droughtmonitor.unl.edu/About/AbouttheData/DroughtClassification.aspx.

[11] R. Walton, "NERC assessment identifies new risk to grid reliability: energy policy," Utility Dive, 23 August 2023. [Online]. Available: https://www.utilitydive.com/news/nerc-assessment-new-risk-grid-reliability-energy-policy/691590/.

[12] International Energy Agency, "The energy world is set to change significantly by 2030, based on today's policy settings alone," International Energy Agency, 24 October 2023. [Online]. Available: https://www.iea.org/news/the-energy-world-is-set-to-change-significantly-by-2030-based-on-todays-policy-settings-alone.

[13] T. Karpavicius, T. Balezentis and D. Streimikiene, "Energy Security Indicators for sustainable energy development: Application to electricity sector in the context of state economic decisions," *Sustainable Development*, vol. 33, no. 1, 2024.

[14] S. Adomako and M. D. Tran, "Scaling up sustainable innovation: Stakeholder ties, eco-product innovation, and new product performance," Sustainable Development, vol. 32, no. 1, pp. 624-634, 2024.

[15] G. Siciliano, L. Wallbott, F. Urban, A. N. Dang and M. Lederer, "Low-carbon energy, sustainable development, and justice: Towards a just energy transition for the society and the environment," *Sustainable Development*, vol. 29, no. 6, 2021.

[16] FERC.

[17] United Nations, "High-level political forum on sustainable development: Progress on the Sustainable Development Goals," 2023.
[18] T. Silvestri, "Complex System Reliability, A Graph Theory Approach," 2014.
[19] NERC, "Adequate Level of Reliability Definition (Informational Filing)," 2013.
[20] NREL, "Explained: Fundamentals of Power Grid Reliability and Clean Electricity," 2024.
[21] International Energy Agency, "Empowering Ukraine Through a Decentralized Electricity System," IEA Publications, 2024.
[22] UN Human Rights Monitroing Mission in Ukraine, "Attacks on Ukraine's Energy Infrastructure: Harm to the Civilian Population," 2024.
[23] Midwest Reliability Corporation, "Regional Risk Assessment, January 2025," 2025.
[24] Midwest Reliability Organization, "Uncertain Energy Availability," 2025, [Online]. Available: https://www.mro.net/organizational-groups/regional-risk/uncertain-energy-availability/.
[25] M. Walton, "Energy has a role to play in achieving universal access to clean water and sanitation," 2018. [Online]. Available: https://www.iea.org/commentaries/energy-has-a-role-to-play-in-achieving-universal-access-to-clean-water-and-sanitation.
[26] IEA, IRENA, UNSD, World Bank, WHO, "Tracking SDG 7: The Energy Progress Report," World Bank, Washington, DC, 2024.
[27] M. T. E. Griffith Jr., "Strategic Attack of National Electric Systems," Air University Press, Maxwell AFB, Alabama, 1994.
[28] N. Kostyuk and Y. M. Zhukov, "Invisible Digital Front: Can Cyber Attacks Shape Battlefield Events?," Journal of Conflict Resolution, 2017.
[29] Human Rights Watch, "The Rwandan Genocide: How It Was Prepared," 2006.
[30] Council on Foreign Relations (CFR), "Conflict in the Democratic Republic of Congo," Center for Preventative Action, June 2025. [Online]. Available: https://www.cfr.org/global-conflict-tracker/conflict/violence-democratic-republic-congo#:~:text=In%20early%202025%2C%20fighting%20between,humanitarian%20crises%20in%20the%20world..
[31] Council on Foreign Relations, "Conflict in the Central African Republic," Center for Preventative Action, 2024. [Online]. Available: https://www.cfr.org/global-conflict-tracker/conflict/violence-central-african-republic.
[32] Silicon Valley Power, "Appliance Energy Use Chart," City of Santa Clara, [Online]. Available: https://www.siliconvalleypower.com/residents/save-energy/appliance-energy-use-chart#:~:text=ENERGY%20STAR%20Refrigerator%20(side%20by,ft.

CHAPTER

3

Fragility, Geopolitical and Security Risks, and Infrastructure Interdependencies

Abstract

This chapter examines the external operating environments that shape electric grid reliability, emphasizing how societal stability, infrastructure interdependencies, and security conditions create powerful forces that either support or undermine grid performance. Building on empirical relationships between state fragility indicators and quality of electricity supply, the chapter demonstrates that reliability outcomes are deeply intertwined with these societal conditions, a connection which remains underrepresented. The analysis further illustrates how critical infrastructures—communications, water systems, fuel networks, transportation, and finance—act as a system-of-systems whose collective integrity determines the functional reliability of the electric grid. Finally, the chapter evaluates the evolving threat environment, including the growing convergence of cyber and physical risks and the increasing geopolitical weaponization of energy. Through these interconnected domains, this chapter shows why reliability planning requires not only technical competence, but also strategic awareness of the social, economic, and geopolitical landscape in which electric systems operate.

3.1 Foreword

In 1995, while serving with the United Nations Assistance Mission for Rwanda (UNAMIR) in Kigali, I received reports of an attack on electrical infrastructure near Bujumbura, Burundi, which disrupted power to a water treatment facility.

DOI: 10.1201/9788743813491-3

Based on prior experience with infrastructure failures and public health, we warned colleagues that a disease outbreak was likely. Within days, a cholera outbreak was reported in Burundi. While the specific causal chain was never formally documented, the episode reinforced the well-established relationship between electricity, water treatment, and public health.

A 2023 peer-reviewed study published in Environmental Science & Technology *examined the public-health impacts of extreme heat combined with electrical grid failures. Using Phoenix, Arizona as a case study, the authors modeled a severe heat-wave scenario combined with a multi-day power outage and staged restoration. The analysis found that such a compound event could lead to* approximately ten thousand or more deaths and several *hundred thousand heat-related emergency department visits, highlighting the potentially catastrophic consequences of prolonged power loss during extreme heat.*

Electricity is how we create and distribute prosperity. It is also how we protect our populations: electricity is used to purify water, and keeps our homes and businesses heated or cooled within a comfortable range. Electricity is used to keep our food cold as it travels towards us along the supply chain, keeping it fresh and nutritious while reducing the need for preservatives like salt, which before widespread refrigeration was a major cause of stomach cancer.

Anyone who has lost a freezer full of food during a power outage knows that electricity is only useful to us if its supply is reliable, though. Our transmission and distribution systems must work when the weather is too cold or too hot. They must work in an environment where thieves try to strip substations of copper, and when extremists or adversaries try to make communities go dark by damaging or destroying critical electrical components.

We ensure the reliability of electricity with a combination of resilience and security measures. A lot of work goes into solving this complex problem that protects our families, friends, and communities, but without it there is no prosperity.

Ross Johnson, Bridgehead Security, Electricity Canada

3.2 Societal Stability

3.2.1 Introduction

Volatile, conflict-ridden societies face massive challenges in addressing reliability. In the chapter foreword, Ross brings in a salient context for review: the 1994 genocide against the Tutsi in Rwanda [1], which resulted in the death of an estimated 800,000 Rwandans, mostly but not exclusively Tutsi. This devastating conflict began with a socially complex and multi-generational history rooted in colonial narratives of ethnic hegemony and existential self-defense. The conflict and its effects were exacerbated by economic forces and resource insecurity, with the socioeconomic devastation being inextricably linked to the lasting effects on the sociotechnical system.

For example, a Yemen-focused think-tank (the Sana'a center for strategic studies) wrote that the inability of the non-Houthi government to provide electricity is "one of the most visceral signs of state collapse" in Yemen [2]. The previous chapter explored the relationship between economic conditions and quality of electricity and found that Yemen was a particularly strong outlier in extreme crisis when it comes to electricity, with minimum annual household electricity usages costing Yemenis more than 9 times the annual GDP per capita.

From a grid-centered perspective, societal stability can be described as the level of constancy and equilibrium surrounding the electric grid, its components and sub-systems, and all their interactions and dependencies. The effects of a highly fragile and instable environment on the quality of electricity supply are palpable and are relevant to researchers, policymakers and planners concerned about the reliability of the electric grid. We face an impressive challenge in such situations—improving or maintaining reliability in the midst of a forceful feedback loop between degraded infrastructure and societal instability requires disruption of the feedback loop, in which the effects of instability prevent any lasting improvements to the situation.

The broad impact of severe instability makes disrupting this feedback loop challenging. Conflicts in Rwanda (with several occurring between 1950 and 2000) or conflicts throughout the greater East Central Africa region [3] [4] created lasting impacts across many domains.

Only incremental change addressing all internal and external factors contributing to reliability issues can begin to impart powerful advancements, like lifting a house inch-by-inch to fix the foundation.

3.2.2 Metrics for quantifying

The impact examples indicate an interconnectedness between many of the external factors that have significant bearing on reliability. The primary challenge for grid planners operating in an unstable context is that resolving the technical outputs will not protect the system from the impacts of instability. This section focuses on the use of the state fragility index as an analytical starting point in comparison with the quality of electricity supply factor from the World Bank. The primary findings indicate that stable societies (those with a low fragility index) tend to have a higher quality of electricity supply. Further, the structured index, with many sub-indicators, provides a starting point for understanding key areas of improvement needed to gain ground on the reliability forefront given the complexity of operating in a highly fragile or unstable environment.

3.2.2.1 Background of the state fragility index

The fragile states index (FSI)[1] is a single variable that takes into account a number of active societal pressures. It was developed and is maintained by the Fund For Peace (FFP) and uses a mixed qualitative and quantitative approach that borrows the indicators from the Conflict Assessment System Tool (CAST)[2], also developed by the FFP.

A comparison of the state fragility indices and quality of electricity supply ratings for 136 countries reveal a moderate correlation between the two (R^2 value of 67%); this correlation is only slightly strengthened by the removal of cohesion indicators (security apparatus, factionalized elites, group grievance) and social indicators (demographic pressures, refugees and IDPs). Unfortunately, this data contains gaps that hinder further analysis—one example being the unavailability of quality of electricity data for some of the most “highly fragile” states, as illustrated in Figure 3.1. The data gaps themselves nevertheless reveal an interesting undercurrent: data collection, including data regarding the reliability of an electric grid, is highly challenging in fragile states, which often leads to critical knowledge deficiencies and hinders the development and implementation of adequate solutions. However, even when data is difficult to collect at a high-level, policymakers and planners within a fragile or semi-fragile context are likely attuned to the operational disruptions and quality of electricity in their system, even without access [5] to comparative data benchmarking with other systems.

[1] For more information on the FSI see: https://fragilestatesindex.org/methodology/

[2] For more information on CAST see: https://fundforpeace.org/2014/03/10/cast-conflict-assessment-framework-manual-2014-reprint/

Figure 3.1 Quality of electricity supply data highlights data gaps in highly fragile states.

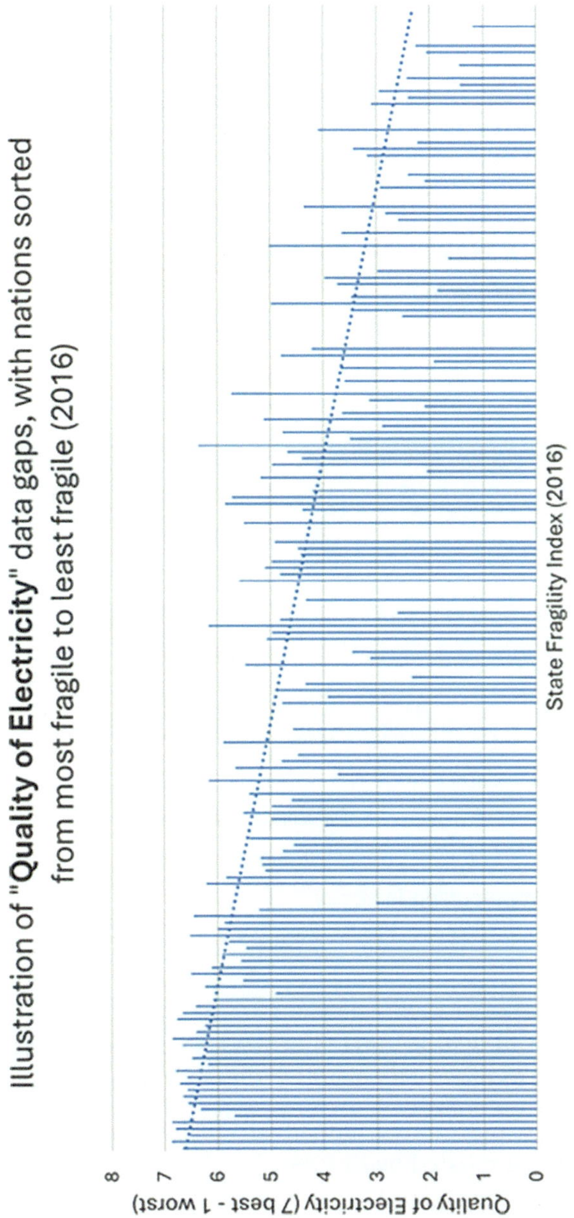

The correlations between the individual indicators within the FSI and the quality of electricity supply are summarized in Table 3.1. The indicators are ranked from 1 to 12 from best performing fit with quality of electricity supply (highest R^2) to worst performing fit (lowest R^2).

Table 3.1 Table of fragile state indicators and their goodness of fit with quality of electricity supply in countries across the globe indicate that certain dimensions of fragility align more closely with quality of electricity supply and reliability.

Fragile states indicators' R^2 fit with quality of electricity supply	
FSI indicator (best–worst rank 1–12)	R^2 value
Cohesion indicators	
Security apparatus (7)	0.4742
Factionalized elites (12)	0.0064
Group grievance (11)	0.1378
Economic indicators	
Human flight and brain drain (5)	0.5047
Economic inequality (3)	0.6496
Economic decline (4)	0.6218
Political indicators	
State legitimacy (8)	0.4048
Public services (1)	0.7673
Human rights and rule of law (10)	0.3672
Social indicators	
Demographic pressures (2)	0.6733
Refugees and IDPS (9)	0.4003
Cross-cutting Indicator	
External intervention (6)	0.4799
Overall FSI value	0.6742

The highest performing individual indicators were *public services, demographic pressures, economic inequality, economic decline*, and *human flight and brain drain.*

For the highest performing individual FSI indicator—*public services*—the key characteristics are listed by FFP as:

- Access to housing
- Housing costs

- Education enrollment
- Literacy rates
- Fuel supply
- Access to medicine
- Number of clinics or hospitals
- Number of physicians
- Infant mortality
- Potable water
- Public services equality
- General conditions of public services
- Roads
- Sanitation
- Airports
- Railroads.

A modified compound indicator using only the top 5 individual indicator values for the 136 countries compared to the quality of electricity supplied produces an R^2 value of 0.7836 (Figure 3.2).

Figure 3.2 Quality of electricity supply plotted against a modified fragility index showing a moderate relationship between the variables.

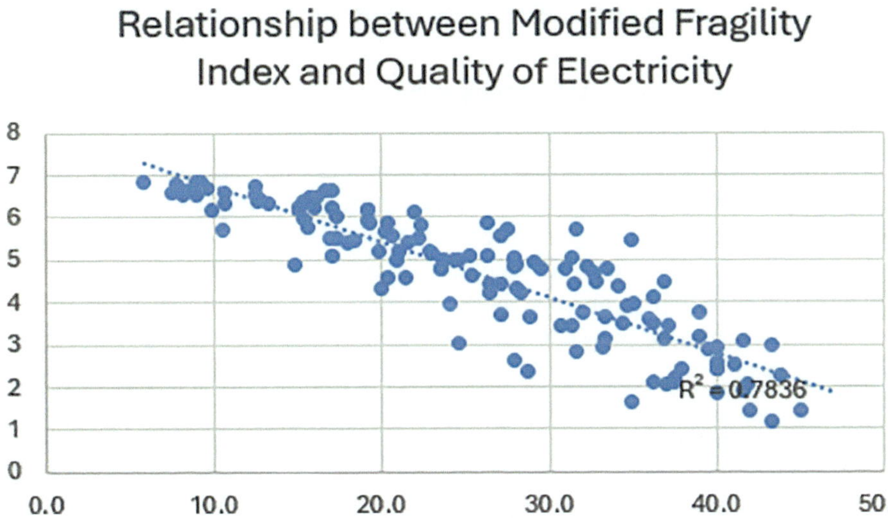

Although public services (general infrastructure and resource availability) is a strong individual indicator, the performance of the modified FSI suggests that the other factors taken into account may help to further explain variability in the data.

Another takeaway from the above analysis confirms an intuitive fact. One of the strongest indicators for stability is the state's economic situation, which has a substantial bearing on the electricity sector since it's adequacy requires significant, consistent long-term investments to meet continued demand growth. Destabilizing societal events, including economic disruptions have the potential to disrupt dynamics of the electric grid, and potential subsequent cascading effects on the many interdependent systems that enable the functioning operation of the electric grid. Tracing the effects between external events and internal impacts is essential for maintaining grid reliability.

In many of these environments, lasting resource insecurity and longstanding economic issues are intermingled with emerging and recurring conflicts, and the state experiences broad impacts on the sociotechnical environment, including the electric grid. Within the FSI framework, resource pressures were identified as an issue contributing to instability. These resource pressures include energy resource pressures and other infrastructures (such as communications, water, etc.) that contribute positively to electric grid reliability. Given that conflicts can both be exacerbated by resource pressures, and contribute to resource pressures, this is identified as a critical feedback loop that in many pre-conflict and post-conflict societies becomes locked-in. Disrupting the loop or assuming it as a permanent condition and developing security and resilience measures as an adaptation become the paths towards a reliable electricity system for these contexts.

3.3 Undercurrent: Infrastructure Interdependencies

Electric power system reliability does not exist in isolation. Beneath the visible technical operations lies an undercurrent of interdependent infrastructures, including transportation, communications, water, fuel supply, and finance. These systems collectively determine the resilience and functionality of the electric grid. When one of these enabling infrastructures experiences disruption, cascading effects can occur that amplify both technical and societal consequences.

3.3.1 Mutual dependencies

Electricity serves as the foundation for all other critical infrastructure sectors, yet the electricity sector itself depends on those same systems for continued operation.

Communications. Modern grid management relies extensively on telecommunications for supervisory control and data acquisition (SCADA), dispatch coordination, and real-time situational awareness. Interruptions in communications networks can delay fault detection, isolation, and restoration, transforming otherwise manageable disturbances into prolonged outages.

Water systems. Many generating units depend on water for cooling, while municipal water treatment and distribution systems depend on electricity to operate pumps and purification processes. Failures in either system can propagate rapidly to the other, compounding both operational and public health risks.

Financial systems. Utility operations are sustained through continuous financial transactions, including billing, payroll, and procurement. These activities depend on the stability of both power and telecommunications networks. Extended outages can therefore undermine the financial integrity of utilities and create a self-reinforcing decline in service quality.

Fuel and supply chains. Power generation and restoration activities require consistent access to fuel, spare parts, and maintenance materials. During crises, disruptions to transport routes or shortages of critical supplies can immobilize maintenance operations and extend recovery timelines. These conditions are particularly common in fragile or conflict-affected states, where infrastructure and logistics systems are already under significant strain.

Assets, people, materials, technologies can become critical constraints under extreme circumstances. A significant takeaway from the COVID-19 pandemic across many sectors was our societal dependence on the continuous movement of goods and services; the pandemic illustrated how unexpected conditions can quickly have cascading impacts across industries. Shockwaves rippled through world economies and major supply shortages produced significant constraints for dependent outputs.

Mirroring the importance of fuel supply to the adequacy of the generation fleet, other inputs to the system such as material goods and services, the flow of technology, and information, and the human resources that enable the effective operation of the electric grid are critical inputs that face increasingly relevant risks across a variety of contexts.

Through these vehicles, the North American electric sector faces exposure to geopolitical risks, political decisions related to tariffs or supply/demand manipulation, and sensitivity to distant markets; mitigations are typically oriented towards bolstering domestic production. However, underdeveloped or developing countries and their electric grids face a distinct set of constraints; including inability or unwillingness to invest in new technologies, or lack of trained personnel to staff a highly reliable grid.

Coupling this with critical dependencies on a small number of producers, and exposure to risk related to procurement of large power transformers, copper wire or electronics used by the electric sector, create a daunting picture for developing nations.

These constraints are not static and can be stochastically modeled as distributions of delay or non-availability based on empirical supply chain indicators. In resilience planning, the temporal availability of replacement parts or critical infrastructure components can be integrated into probabilistic simulations as random lag variables, increasing realism and decision-usefulness of reliability models. Scenario planners may consider incorporating leading indicators of supply chain stress such as global shipment times, export controls, or vendor insolvency risk, into their asset-level recovery and availability projections. Doing so allows grid planners to more accurately simulate and prepare for the full delay arc associated with contingency events, not just the initiating disruption itself.

Energy availability is an emerging pillar of reliability. The system of systems approach for thinking about the electric grid's reliability quickly brings to the forefront a key contribution to risks to reliability faced by the electric grid: uncertain *energy availability*. In the Midwest Reliability Organization's Regional Risk Assessment for 2025 [5], "uncertain energy availability" is identified as the most important emerging risk to reliability as its impacts can be severe and it's likelihood is high. Although "uncertain energy availability" encompasses a number of factors that are traditionally addressed under "resource adequacy"—a traditional pillar of reliability, MROs own argument suggests that the compilation of multiple external drivers have now necessitated a broadly expanded view of reliability risk, stating, "because of these factors, a new way of understanding what energy is available... to meet demand at all times must be developed." [6]. From a system-internal perspective, interconnection queues, the changing resource mix (installed capacity), increased electrification (demand) and the added volatility of large sink-point loads (such as data centers) create a complex operating environment—but one that is perhaps adequately ideated by traditional resource adequacy approaches. However, industry is in strong concordance that a new approach is urgently needed. The ability to ensure that externally supplied resources are always available to meet demand requires an expansion of the field of analysis to consider key driving (external factors) under a new pillar: *energy availability*. Some of the key factors influencing this pillar include weather and climate, energy policy at all levels of governance, natural resource supply, material supply chain and manufacturing conditions, and more.

Issues related to adequacy are deeply interlinked with energy availability. Seasonal studies by FERC, NERC, and the NERC regions suggest that the weather outlooks during peak seasonal loads, such as the hottest point of summer and coldest days in deep winter, are a driving factor in the determination of how energy markets will perform, and equally so, how the electric grid will fare. This interdependency is especially forceful for the natural gas inputs to the electric grid. Within the context of the North American grid, it's impossible to ignore the importance of and growing dependence on natural gas. This is not a value statement or a presumption of good or bad; it is simply a fact that innovations in the natural gas extraction process and market mechanisms have lowered the cost of electricity to some consumers by increasing the sector's reliance on natural gas. Reliability organizations in North America have made clear that fuel unavailability is a major concern for the electric grid. This is not a new concept, and it is really quite intuitive that the inputs to the sector must be secured for the sector's outputs to be secure. However, *energy availability* as an independent factor remains on the margins.

3.3.2 Examples of the impact of energy availability on electric reliability

Chronic fuel shortages and societal instability compounding impacts of inadequate resources. An extreme case that illustrates a multi-faceted crisis of energy and economy; Lebanon has experienced severe challenges to the reliable production of electricity since the Lebanese Civil War (1975–1990). Before the civil war, Lebanon "produced enough electricity to export surplus power", but the destruction of the small country's energy and electricity infrastructure during the 15-year civil war, in addition to issues of mismanagement, economic issues, and more, led to the current challenge faced by Lebanon: fuel supplies are not able to meet power sector demands, resulting in ongoing blackouts [7].

Droughts in Venezuela contribute to collapsing electric grid [8]. During the early 2020s, Venezuela suffered a total blackout across the country under normal load conditions, lasting several days, with rolling blackouts remaining a consistent feature of life for Venezuela. Structural conditions that contributed to this collapse included a lack of diversity in the energy mix and overreliance on a handful of generators representing single points of failure in the generation and transmission subsystems. The structural issues combined with circumstances related to severe drought conditions which reduced the energy output of the primary generation source (hydropower) highlight the importance of energy availability. Securing a consistent energy supply is of critical importance, as is analyzing the structural conditions that can lead to significant energy vulnerabilities, such as overreliances on one fuel source.

Rolling blackouts in Texas during Winter Storm Uri. During February 2021, Texas faced temperatures much colder than ever before; a series of failures resulted in a feedback loop between loss of electricity and loss of energy supply. From NREL's case study on the event:

"Although there were many drivers of the shortfall, including inoperable generating facilities of all types because of the cold weather, one contributor was a lack of natural gas supply due to frozen gas wellheads and electricity disruptions at gas extraction and network points. The Texas power system operator (the Electric Reliability Council of Texas, or ERCOT) estimated that during the hour of highest outages, as much as 6 GW of generation—approximately 12% of the total outages at the time—was offline due to an inability to obtain natural gas supplies" [10].

3.3.3 Cascading and feedback effects

The interactions among critical infrastructures are complex and nonlinear. A disruption in one domain may initially appear localized but can quickly spread through hidden couplings such as fuel delivery schedules, shared communication dependencies, or regional logistics hubs. Amplifying feedback loops were identified in the previous section, including in 2021 in Texas during Winter Storm Uri [6], and around the same time across multiple sectors during the Venezuelan collapse [7].

In fragile governance contexts, these interconnections often lack redundancy and coordination, resulting in conditions where even small shocks escalate into systemic crises.

Such interdependencies also interact with social and economic dynamics. Declining reliability can erode public confidence, deter investment, and contribute to broader social instability. The resulting feedback loop, in which degraded infrastructure exacerbates societal fragility and vice versa, poses a significant challenge for planners and policymakers seeking to improve reliability under constrained conditions.

3.3.4 System-of-systems planning

Recognizing these interdependencies requires a shift from single-sector reliability analysis toward a system-of-systems framework. Reliability planners are increasingly collaborating with adjacent sectors such as telecommunications, transportation, and emergency management to identify shared vulnerabilities and restoration priorities. Analytical tools including interdependency matrices, network models, and consequence-based frameworks (such as CIRA or CIMS) are being used to map and quantify how disruptions in one infrastructure can propagate across others.

3.3.5 Implications for fragile and transitional states

In fragile or transitional environments, the management of infrastructure interdependencies can either accelerate recovery or deepen systemic collapse. When governments and utilities coordinate restoration hierarchies that prioritize electricity supply for water treatment, fuel logistics, healthcare, and essential communications, the overall system stabilizes more rapidly. Conversely, where such prioritization is absent, cascading failures across dependent infrastructures exacerbate humanitarian and economic crises.

Consequently, resilience investment should not be limited to the physical hardening of grid assets. It should also strengthen the connective mechanisms that link the electricity sector to its interdependent systems, including logistics coordination, shared communication protocols, and multi-sector governance structures. By addressing these interdependencies as integral components of reliability planning, long-term system resilience be achieved.

3.4 Security and geopolitical risks

Mitigating risks related to cyber and physical threats, vulnerabilities, and consequences is increasingly important for electric grid reliability. The threat of physical attacks against electricity infrastructure during war-time date back to the 1930s [4], and the first instance of cyber activities deployed during war has been documented in the ongoing 2010s–2020s Ukraine–Russia conflict [5]. However, the ongoing threat from criminals, activists, and other non-state adversaries has been underrepresented within the reliability discussion, despite increasing societal dependence on electricity, blending of threat actors, emergence of crime-as-a-service, and the increased capacity for weaponization of electricity. Adding to these threats are the increasing vulnerabilities of electricity infrastructure: the electric grid's communication and control features have expanded into a highly complex cyber-physical system (CPS), thus increasing the attack surface for a variety of motivated threat actors in both the cyber and physical realms of operation. Finally, the consequences of successful attacks, regardless of motivation or domain (cyber, physical, or blended) have been proven in both wartime and civilian contexts.

3.4.1 The influence of geopolitical risks

One anecdotal example of the direct effect of geopolitical situation on the electric grid is found in the Russia–Ukraine war; in 2024 Ukraine saw systematic attacks on the electricity infrastructure. According to the IEA: "Since Russia's full-scale invasion of Ukraine in February 2022, nearly two-thirds of Ukraine's dispatchable power capacity has been occupied, damaged, or destroyed." [6]. The effects have been devastating; in addition to the loss of human life, the widespread effects and losses will echo through Ukrainian society for years or even decades to come [7]. Russia is one adversary that has readily demonstrated a willingness to weaponize electricity and energy on a variety of wartime fronts, such as in the example of attacking electricity infrastructure throughout the Russia–Ukraine war as a component of warfare, but also as in the example of Russia's curtailment of gas supplies to oppositions in Europe. Even after growing tensions at the start of the Russia–Ukraine conflict in 2014, Europe's strong reliance on Russian energy [8] demonstrated a significant weakness that Putin was ready to leverage by 2021. A number of factors contributed to Europe's broad inability to prepare for geopolitical weaponization of energy in this particular context: overall a decentralized energy policy and strong autonomy among European states contributed to a lack of a coherent energy strategy and a piecemeal approach to policy. Europeans have also strongly advocated for the phasing out of coal and nuclear plants in order to address climate risks, and maintaining reliability and quality of electricity has required the use of LNG as a bridge fuel. Perhaps the most heavily weighed factor was the misconception that mutual economic benefit would weigh in Europe's favor and prevent weaponization of energy supply. All these factors contributed to Europe's, but especially Germany's lack of willingness to address the geopolitical risk associated with strong reliance on Russian LNG until of curtailment of supply began to rear its ugly head.

Noting this anecdote as a recent example, weaponization of energy and electricity is a longstanding tool for foreign policy; much like supply chain issues, resource constriction of critical minerals, and market dominance as issues considered in particular within the lens of tensions with China, energy and electricity dependence should also be consistently considered within the geopolitical context not only by foreign policy analysts, but by reliability planners.

3.4.2 Physical security vulnerabilities and the reliability interface

Physical security is the bridge between societal stability and operational reliability. It represents the point where the external environment directly interacts with the infrastructure of the electric grid. The frequency and severity of physical attacks or intrusions are not random; they are shaped by economic and geopolitical factors, criminal opportunity, activist and extremist movements, and the broader legal environment. When judicial systems are weak or enforcement capacity is low, as in parts of Brazil and other high-crime regions, deterrence becomes ineffective, and opportunistic crimes proliferate. Conversely, societies with functioning legal systems and consistent prosecution of crimes enjoy a measure of societal deterrence that significantly reduces low-motivation threats.

The North American Electric Reliability Corporation (NERC) has identified physical security vulnerabilities as a continuing and evolving risk to bulk power system reliability. These vulnerabilities become reliability issues when events at the physical layer—such as theft or vandalism—trigger operational disturbances or forced outages. To understand the influence of specific vulnerabilities and conditions on the threat environment, it is useful to think in four concentric physical spheres of influence:

- Site level: The most immediate layer concerns facility-specific vulnerabilities. Examples include unfenced perimeters, unhardened fence lines, clear ballistic lines of sight, and unsecured access points. Even small deficiencies in site design can create conditions in which attackers or opportunists can cause measurable operational impacts.
- Area level: This encompasses the immediate surroundings of a facility: the neighborhood, parish, or municipal context. Crime density, policing practices, and socioeconomic conditions such as poverty or homelessness shape the background threat environment. For example, high levels of homelessness or untreated mental illness can increase exposure to property crimes, vandalism, or erratic behavior near critical infrastructure. Research published in the National Library of Medicine found that homeless individuals with mental disorders commit criminal offenses at rates dozens of times higher than the general population, underscoring the connection between social welfare, policing, and the security posture of nearby facilities.
- Regional level: Broader state or provincial conditions, economic health, regulatory rigor, and law enforcement resource allocation, extend the area-level picture. Regional regulation of scrap metal markets and border trade, the prevalence of organized crime or gangs, and common practices of prosecution all influence the overall likelihood of physical incidents targeting electric infrastructure. For instance, unregulated scrap markets have historically correlated with higher rates of copper theft at electric sites.
- National level: Finally, the federal and geopolitical context defines the outer boundary of physical security risk. Legislative frameworks, such as laws governing infrastructure protection and criminal penalties for sabotage, form the deterrence baseline. In parallel, nation-state adversaries or organized transnational groups introduce higher-order risks that merge with cyber or hybrid threats, particularly during geopolitical conflict.

Physical security risk is therefore multidimensional, shaped by both external inputs and internal states. Externally, the threat environment is influenced by variables affecting crime—such as economic confidence, unemployment, and property-crime statistics [13]. Internally, asset vulnerabilities depend on system design, facility layout, and the effectiveness of countermeasures like surveillance, intrusion detection, and rapid response capability.

When external and internal conditions intersect, the consequences become measurable in reliability terms: loss of load, service disruption, or repair and replacement costs. These outcomes should be recognized as reliability events whose probability can be quantified and mitigated like any other operational risk. Notably, some threat actors have demonstrated a deliberate use of electric outages as force multipliers, attacking substations in an attempt to create chaos for secondary crimes. This underscores the societal importance of electricity and the central role that physical security planning plays for critical infrastructure in order to prevent cascading impacts and loss of service.

3.4.3 Cybersecurity and the reliability interface

Cyber risk is where the grid's increasing digitalization meets adversary intent. As SCADA, EMS, and DER management platforms, protection relays, and field devices become tightly coupled through IP-based networks, reliability outcomes depend not only on classic "availability" engineering but also on how well cyber threats are deterred, detected, contained, and recovered. In the last decade, real-world attacks have demonstrated that cyber operations can open breakers, disable protections, blind operators, and delay restoration—i.e., create direct reliability events rather than mere IT nuisances. Ukraine's 2015 and 2016 incidents (and later 2022/2023 variants) remain canonical examples of cyber-physical impacts on power systems.

Similar to the physical domain, it's useful to think about four spheres of influence that translate cybersecurity conditions into reliability risk:

- Site/asset level (OT and substation/plant systems). Device exposure, insecure-by-default services, flat networks, weak access control for IEDs/RTUs/PLCs, and insufficient logging create pathways for adversaries to pivot from IT into EMS/SCADA or to send malicious control commands. Controls mandated by NERC CIP (e.g., asset categorization, electronic access control, incident response, and low-impact obligations) are designed to shrink attack surfaces and raise the bar at the device and enclave level.
- Area/enterprise level (utility IT/OT convergence). Weak segmentation between corporate IT and operations networks, unmanaged vendor remote access, and inconsistent identity governance

often explain how attackers reach control environments. Ransomware in enterprise domains has increasingly disrupted industrial organizations' operations, even when OT was not directly encrypted, by forcing shutdowns during investigation and recovery.

- Regional/ecosystem level (supply chain, partners, ISACs). Grid operators rely on OEMs, integrators, telecoms, and cloud/SaaS providers. Compromise of these partners—firmware, update servers, or remote support portals—can propagate widely. Sector collaboration and near-real-time threat sharing through the E-ISAC aim to reduce dwell time and coordinate mitigations across the interconnection. NERC's CIP-008 reporting via E-ISAC also provides visibility into incident patterns for the industry.
- National/geopolitical level (state actors and policy). Nation-state units have repeatedly targeted energy systems with bespoke ICS-capable malware and living-off-the-land tradecraft to reduce detection and accelerate disruptive effects. Strategic tensions correlate with elevated cyber operations against critical infrastructure. Policy guidance has therefore shifted from pure prevention to cyber-resilience; designing electricity systems to withstand, adapt to, and rapidly recover from cyber incidents while maintaining essential services.

3.4.4 From cyber conditions to reliability consequences

Guidance from CISA and the IEA emphasizes that full prevention is impossible, so operators should prioritize measures that measurably reduce the probability and duration of loss-of-load outcomes: (1) rigorous asset inventory and zone/conduit segmentation; (2) hardened and monitored remote access with MFA and just-in-time privileges; (3) engineering-grade allow-listing and strict change control on IEDs/PLCs; (4) OT-capable monitoring and incident response (with "gold images" and offline backups for rapid rebuild); (5) blackstart/restore drills that assume cyber-degraded conditions; and (6) active participation in E-ISAC sharing and exercises.

3.4.4.1 Analyst's checklist

- **Threat environment:**
    - Current sector advisories (E-ISAC, CISA ICS)
    - Regional ransomware prevalence
    - Known actor TTPs against power systems (e.g., OT-specific malware, LotL).
- **Exposure and controls:**
    - Asset inventory completeness
    - Routable vs. non-routable connectivity
    - Remote-access pathways
    - Segmentation quality
    - CIP control coverage at low/medium/high impact BES Cyber Systems.

- **Response and recovery:**
  - Evidence of OT-specific IR plans (CIP-008)
  - Backup/restore tests for EMS/SCADA and substation HMIs
  - Golden configurations
  - Restore-under-duress drills.
- **Consequence modeling:**
  - Credible cyber-initiated outage scenarios (breaker operations, relay misconfiguration)
  - Restoration timelines
  - Cascading risk to dependent infrastructures.

3.5 Exercises

(1) Disrupting the feedback loop
In fragile or semi-fragile states, degraded infrastructure and societal instability reinforce one another. If a planner could only intervene in one side of the loop first—infrastructure conditions or societal conditions—which would likely produce the most durable long-term reliability gains, and why? Support your argument with reasoning rather than examples alone.

(2) System-of-systems prioritization under constraint
During a prolonged grid disruption, multiple interdependent infrastructures will compete for limited restoration resources (water, telecom, fuel logistics, healthcare, financial systems). How should priorities be established? Propose a principled restoration hierarchy and justify it ethically, economically, and socially—not just technically.

(3) The legitimacy paradox
The chapter suggests that public trust and state legitimacy shape the ability to maintain reliability. But legitimacy itself is often a result of reliable service, not a precondition. How can a utility or operator build legitimacy before reliability has been fully restored? What forms of transparency, communication, or shared governance might shift public perception?

(4) Security as a societal mirror
Physical and cyber threats are partially a reflection of the surrounding social environment—motivation, opportunity, grievance, perceived justice, and governance capacity. If this is true, should reliability planning be understood as a security function, a social investment, or both? Argue for one framing, then critique your own position.

(5) Energy as a tool of statecraft
Energy systems can be weaponized through physical destruction, cyber disruption, or market manipulation. Under what conditions, if any, is it strategically rational for a state to tolerate high energy dependency on a geopolitical rival? Describe thresholds at which interdependence shifts from stabilizing to dangerous.

(6) Designing for inevitable failure
Given that prevention of all attacks—physical or cyber—is impossible, how should reliability planning balance investment between hardening, deterrence, and rapid restoration capacity?

Which of these has the greatest leverage in complex or fragile political environments, and why?

(7) The ethical boundaries of resilience

In some contexts, resilience planning may mean prioritizing service to certain institutions or populations (e.g., hospitals, government agencies, major economic nodes) before others. How should planners navigate the ethical tension between maximizing systemic stability and preserving equity in access to electricity?

References

[1] Human Rights Watch, "The Rwandan Genocide: How It Was Prepared," 2006.

[2] "Blackouts and Blackholes: Yemen's Vanishing Electricity Supply - The Yemen Review, September and October 2023," Sana'a Center For Strategic Studies. https://sanaacenter.org/the-yemen-review/sept-oct-2023/21234

[3] Council on Foreign Relations (CFR), "Conflict in the Democratic Republic of Congo," Center for Preventative Action, June 2025. [Online]. Available: https://www.cfr.org/global-conflict-tracker/conflict/violence-democratic-republic-congo#:~:text=In%20early%202025%2C%20fighting%20between,humanitarian%20crises%20in%20the%20world..

[4] Council on Foreign Relations, "Conflict in the Central African Republic," Center for Preventative Action, 2024. [Online]. Available: https://www.cfr.org/global-conflict-tracker/conflict/violence-central-african-republic.

[5] "Open Knowledge Repository," Worldbank.org, 2024. https://openknowledge.worldbank.org/entities/publication/bb2b7691-4116-5354-91d1-41858347253e

[6] Staff, FERC - NERC - Regional Entity, "The February 2021 Cold Weather Outages in Texas and the South Central United States," 2021.

[7] S. Weir, "Electric Grid Collapse in Venezuela: Conceptualizing Systemic Grid Failure Through a Modified FMEA Risk Assessment," Johns Hopkins University, 2021.

[8] M. T. E. Griffith Jr., "Strategic Attack of National Electric Systems," Air University Press, Maxwell AFB, Alabama, 1994.

[9] N. Kostyuk and Y. M. Zhukov, "Invisible Digital Front: Can Cyber Attacks Shape Battlefield Events?," *Journal of Conflict Resolution,* 2017.

[10] International Energy Agency, "Empowering Ukraine Through a Decentralized Electricity System," IEA Publications, 2024.

[11] UN Human Rights Monitroing Mission in Ukraine, "Attacks on Ukraine's Energy Infrastructure: Harm to the Civilian Population," 2024.

[12] FBI, "UCR: Variables Affecting Crime".

[13] United Nations, "High-level political forum on sustainable development: Progress on the Sustainable Development Goals," 2023.

[14] T. Silvestri, "Complex System Reliability, A Graph Theory Approach," 2014.

[15] NERC, "Adequate Level of Reliability Definition (Informational Filing)," 2013.

[16] NREL, "Explained: Fundamentals of Power Grid Reliability and Clean Electricity," 2024.

[17] V. Sultan and B. Hilton, "Electric grid reliability research," *Energy Informatics,* 2019.

[18] Midwest Reliability Corporation, "Regional Risk Assessment, January 2025," 2025.

[19] Midwest Reliability Organization, "Uncertain Energy Availability," 2025, [Online]. Available: https://www.mro.net/organizational-groups/regional-risk/uncertain-energy-availability/.

[20] M. Walton, "Energy has a role to play in achieving universal access to clean water and sanitation," 2018. [Online]. Available: https://www.iea.org/commentaries/energy-has-a-role-to-play-in-achieving-universal-access-to-clean-water-and-sanitation.

[21] IEA, IRENA, UNSD, World Bank, WHO, "Tracking SDG 7: The Energy Progress Report," World Bank, Washington, DC, 2024.

[22] T. Karpavicius, T. Balezentis and D. Streimikiene, "Energy Security Indicators for sustainable energy development: Application to electricity sector in the context of state economic decisions," *Sustainable Development,* vol. 33, no. 1, 2024.

[23] S. Adomako and M. D. Tran, "Scaling up sustainable innovation: Stakeholder ties, eco-product innovation, and new product performance," *Sustainable Development,* vol. 32, no. 1, pp. 624-634, 2024.

[24] G. Siciliano, L. Wallbott, F. Urban, A. N. Dang and M. Lederer, "Low-carbon energy, sustainable development, and justice: Towards a just energy transition for the society and the environment," *Sustainable Development,* vol. 29, no. 6, 2021.

CHAPTER

4

Conclusions on The Living System

Abstract

This chapter concludes by reframing reliability as a living relationship between technical systems and the societies that sustain them. It argues that reliability cannot be isolated from its context: it mirrors the coherence, stability, and trust within the institutions and communities it serves. Through synthesis of the preceding chapters, the discussion emphasizes that the grid's performance depends not only on physical assets and operating standards, but on governance quality, public confidence, and the alignment of interdependent infrastructures.

The chapter explores adaptation as the defining posture of modern reliability, shifting the focus from control to flexibility amid uncertainty. It highlights trust as an invisible yet indispensable form of infrastructure, connecting operational reliability to civic integrity. Interdependence is presented not as a vulnerability to be eliminated, but as a design principle that enables coordinated recovery across sectors.

Ultimately, reliability emerges as a practice of stewardship rather than perfection; a continuous commitment to foresight, cooperation, and resilience in the face of a changing world. It closes with a call to view the grid as a living system: dynamic, interdependent, and reflective of the societies that depend upon it.

DOI: 10.1201/9788743813491-4

4.1 Introduction

Reliability has long been defined as a condition—a measurable state of adequacy, stability, and performance. But in truth, reliability is not a static property of machines; it is a living relationship between infrastructure and the society that sustains it. The grid breathes with the same rhythm as the world around it, reflecting its strengths, tensions, and contradictions. When societies grow stable and cohesive, the grid becomes a mirror of that order: predictable, self-correcting, and trusted. When they grow fragmented, under-resourced, or politically brittle, the grid reveals that fragility. Reliability, in this broader sense, is not simply a product of engineering discipline, it is an expression of social coherence.

This idea might seem abstract until one looks closely at what failure reveals. When a grid collapses under pressure—whether from weather, economic crisis, or sabotage—the technical sequence of events is rarely the whole story. Each event exposes decisions made years earlier about maintenance priorities, governance stability, and the balance between cost and resilience. A system does not fail only when a breaker trips; it fails when its enabling environment erodes.

Consider the extended blackouts in Puerto Rico following Hurricane Maria. The physical damage was immense, but the prolonged recovery owed as much to institutional fragmentation and resource inequity as to the storm itself. The event showed how technical reliability collapses when political, financial, and logistical systems lose alignment. That case is not unique—it simply made visible what is usually invisible: that electricity is both an engineering achievement and a social contract.

This book began with that recognition. It argued that reliability cannot be separated from the conditions surrounding it. Each chapter traced a dimension of that environment—economic, institutional, physical, and geopolitical—to show how they converge in determining whether the grid can perform as intended.

The first chapter placed reliability in human context: electricity as a prerequisite for dignity and stability. It reminded us that access without reliability is only a temporary gift—that light which flickers unpredictably is not a symbol of progress, but a reminder of fragility.

The second chapter examined affordability and governance, questioning the common assumption that reliability and affordability are opposites. Through comparative evidence, it showed that systems built on coherent institutions and clear accountability achieve both. What undermines reliability is not the pursuit of affordability itself, but the instability of the frameworks through which affordability is pursued.

The third chapter extended outward into the environments that surround the grid: social stability, interdependent infrastructures, and geopolitical conditions. It drew on risk models and societal fragility indices to demonstrate that reliability outcomes correlate with the strength of governance, public trust, and inter-sector cooperation. In that view, the grid becomes a barometer of societal health, responding to pressures well beyond its physical footprint.

This final chapter gathers those threads. It asks: if reliability is inseparable from its context, how should that change the way we understand and pursue it?

4.2 From Control to Adaptation

The traditional mindset of reliability planning assumes that disturbances are deviations from the norm and that the task is to prevent, isolate, or restore from them. Yet the world that the grid inhabits is shifting faster than that logic allows: climate volatility, digitization, and global interdependence have made disruption a permanent feature, requiring a different posture. Reliability must evolve from the pursuit of control to the cultivation of adaptation. This shift mirrors the evolution of medicine from prevention to resilience: from eradicating disease to managing it sustainably.

Adaptation also demands humility. No model or standard can capture the full complexity of interacting systems, but planners can design for flexibility—modular networks, diversified supply chains, multi-layered restoration strategies, and governance processes that reward learning rather than perfection. The future of reliability will belong to those who treat uncertainty not as an anomaly, but as a design parameter.

4.3 The Central Role of Trust

Underlying every form of reliability is an invisible infrastructure: trust. Operators must trust their data, institutions must trust their partners, and the public must trust that the system will deliver what it promises. Without that trust, even technically reliable systems become politically or socially unstable. This is particularly evident in the growing intersection between energy and information. As the grid becomes increasingly digitized, its reliability will depend as much on cyber hygiene, transparency, and cooperative intelligence as on transformers and turbines. Every kilowatt delivered carries with it an implicit expectation that the system behind it is both competent and honest. Maintaining that expectation is now part of the reliability mission. Trust also links the macro and micro levels of reliability. The same qualities that keep a power pool stable: transparency, coordination, accountability, keep societies stable as well. Reliability, therefore, is not only an engineering outcome.

4.4 Interdependence as a Design Principle

Perhaps the clearest lesson from the preceding chapters is that the grid cannot be isolated from the infrastructures that sustain it. Electricity depends on fuel, water, transportation, communication, and finance—all of which depend in turn on electricity. This circularity makes interdependence both a risk and a resource. The task for reliability planning is not to break these links, but to understand and manage them. Systems designed with awareness of their dependencies can anticipate cascading failures and orchestrate coordinated recovery. The most resilient architectures will be those that treat interdependence as a design principle, building redundancy not only within sectors but between them.

In this sense, reliability planning becomes an exercise in system stewardship, extending beyond megawatts and maintenance cycles into relationships—between institutions, sectors, and even nations. Cooperation, once treated as a policy aspiration, has become an operational necessity.

4.5 The Human Continuum

At its deepest level, reliability is about continuity—the ability of people and institutions to plan their lives with confidence that tomorrow will resemble today. When electricity is reliable, that continuity feels invisible, ordinary. When it disappears, so does the illusion of order. The professionals who work in reliability are profoundly important servants of the public who safeguard more than the flow of current; they preserve the continuity of expectation on which modern life rests and are the custodians of predictability, maintaining the quiet assurance that the tools of progress will respond when called upon.

4.6 Looking Ahead

As we look toward the coming decades, the landscape of reliability will grow more complex. Decarbonization, electrification, distributed generation, and automation will redefine both the grid's structure and its risk profile. The boundaries between electricity, transportation, communications, and data will blur further. Yet the principles that sustain reliability will remain the same: foresight, coordination, transparency, and trust. To meet that future, reliability planning must expand its field of vision. It must integrate the social and the technical, the local and the global, the immediate and the long-term. The goal is not to design a system immune to failure, but to ensure that even in disruption, the foundations of recovery remain intact.

4.7 Conclusion: The Work Still Ahead

Reliability is not a finish line; it is a practice of attention. It asks us to see the grid as an organism that responds to its environment, learns from stress, and mirrors the condition of the society it powers. The more faithfully we understand that mirror, the better we can protect what it reflects.

The work of reliability will never be done, and that is precisely what makes it worth doing. Each generation inherits the same promise: to hand forward a system that is not perfect, but prepared; not invulnerable, but resilient. The measure of success will not be whether the grid never fails, but whether it continues to bind together the fragile and remarkable continuity of modern life.

That is the living system. And that, above all, is the work still ahead.

Index